50¢

CATALOG OF FOOD

Also by Jeffrey Feinman:

CATALOG OF KITS
CATALOG OF FREE THINGS

CATALOG OF FOOD

JEFFREY FEINMAN
ILLUSTRATIONS BY IVOR PARRY

DOLPHIN BOOKS
DOUBLEDAY & COMPANY, INC., GARDEN CITY, NEW YORK
1977

Dolphin Books
Doubleday & Company, Inc.

ISBN: 0-385-11638-1
Library of Congress Catalog Card Number 76–42421
Copyright © 1977 by Jeffrey Feinman
ALL RIGHTS RESERVED
PRINTED IN THE UNITED STATES OF AMERICA
FIRST EDITION

To Rick McCabe,
my favorite food marketer

. . . And special thanks to
Pat Filley and Lindy Hess
for their assistance

CONTENTS

INTRODUCTION	xi
KITCHEN UTENSILS	1
MEAT, FISH AND POULTRY	25
FRUITS AND VEGETABLES	51
CHEESE	63
CONDIMENTS, SPICES, AND SYRUPS	79
CONFECTIONS	101
BEVERAGES: COFFEES AND TEAS	135
HEALTH AND ORGANIC FOODS	151
GROW IT YOURSELF	167
INTERNATIONAL GROCERIES	179

SPECIAL FOODS   191

FOOD KITS   205

INTRODUCTION

Do you remember the taste of real tub butter? Or bread before it was mostly air? Have you ever tasted corncob-smoked and specially aged Vermont Cheddar cheese? And how about the aroma and taste of imported, unblended coffee beans which you have freshly ground yourself? Perhaps you think the days are long past when you can recapture these epicurean delights.

If all that you eat is bought in a supermarket, yes, the taste of the best in foods is probably just a dim, receding memory. Unfortunately, all too many foods are processed to the point that you can hardly discern what you are eating.

But gourmets and gourmands can take heart. The kinds of foods described above—along with succulent seafoods; tender aged beef; ripe, juicy, outsize fruits; smoked meats, poultry, and game; homemade and hand-dipped, creamy mints, chocolates, and other candy treats; tempting appetizers, entrees, and desserts—can grace every table. And it all will be delivered fresh, by your mailman.

Food is the largest single item in the American family budget. It is both a source of concern and a sensual pleasure. The simple fact is that food does not taste as good as it used to because the true natural elements are decreasing as the chemical content is increased.

American marketers, always eager to find a new opportunity, have solved the problem of providing you with truly delectable foods—the fresh, wonderful delicacies you had as a child. As astounding as it sounds, to get good coffee in New York, one must order it by mail; to get real tea (that's not brackish or in dust), one must order it by mail.

WHY "A CATALOG OF FOODS"

Americans are becoming more and more conscious of foods (and what goes into them). And they long, more and more, for gourmet foods and savory favorites they knew in their old neighborhoods and their family kitchens. In most cases, these foods are just not available on the shelves of your grocery store . . . or in the meat section . . . or in the produce section . . . or in the dairy section. Most often, one settles for the ordinary; but there is so much more available. America abounds with special fare which would delight even the most sophisticated palate; and it is yours for a postage stamp. This catalog will give you an idea of all the foods available and where you can buy them. If you have a craving for anything that is out of season or indigenous to a specific region, all you have to do is consult this catalog and your appetite can be satiated by mail.

HOW TO USE THIS BOOK

This book is arranged by food categories; for example Meats, Fish, and Poultry; Condiments, Spices, and Syrups; Confections; etc. You can, therefore, browse through to find your areas of specific interest. (And don't be surprised if they all interest you!) After many of the companies, you will find a paragraph which will provide you with some specific background, including whether or not they have a catalog or information sheets for which you can send. In addition, you will find these backgrounds cross-referenced in other chapters. At the end of many chapters, you will find an "Other Sources" section which provides you with more places to write about a specific category of food.

HOW TO ORDER

Follow the prices in this book for *guidelines only. Please don't order from this book.* The CATALOG OF FOOD took over a year to compile and edit. And in today's inflationary economy, you can't expect these prices to remain stable. *So before ordering anything, write for complete details.*

Where a catalog is mentioned, it's a good idea to write for the catalog first. If a charge is indicated, be sure to include it. It becomes prohibitively expensive for many small manufacturers to send catalogs without charge. (And many will refund the price with your first purchase.)

One last note on ordering. Many of the firms listed are exclusively mail-order ventures. Because of seasonal products, some only operate on a part-time basis. Therefore, if you live near an address listed, don't drop by unless you call first!

WHO IS INCLUDED

Almost anyone who sells food or food-related items by mail is included in this catalog. To compile it, we sent out letters to thousands of companies, made hundreds of phone calls and follow-up efforts. The company names were derived by researchers who pored over newspapers and magazines, talked to trade associations, mail-order firms, and, of course, lots of friends who had ordered these products.

We also ordered and tasted hundreds of these items. (You're right, we spent weeks dieting as a result!) But, in some cases, we had to rely on the descriptions supplied by the growers and manufacturers. If you think this catalog doesn't give you enough information on a particular food or product in which you are interested, don't hesitate to write for further details. Mail-order companies are pros at handling questions, and will be delighted to help you. Finally, nobody paid to be in this catalog. People who receive a good number of mentions got them because of their variety of products, and because they deserved to be mentioned.

WHO IS NOT INCLUDED

If your favorite company, or a company you have heard of, is not listed, it's probably because: 1. We didn't find it. 2. It didn't answer our inquiry or provide proper information in time to meet the deadline for this book.   3. The items were widely available at retail.   4. The firm asked to be deleted.   5. The products were common or unacceptable.

ABOUT MAIL ORDER

Many of the firms here couldn't survive in any other way but mail order. Therefore, their aim is to make you a satisfied customer so that you will want to reorder from them over the years and will be anxious to try other products which they might offer to you. Many companies have been in business from fifty to a hundred years; they couldn't have survived that long without happy buyers.

ABOUT LISTINGS

This book is intended to be a guide. We have used sources and information which we believe to be reliable, but, we cannot guarantee any source, the address, or the information. Listing in this book is not to be considered an endorsement. Conversely, exclusion is not to be considered a rejection.

NOW BEGIN

There's a whole world of exciting food awaiting you in the pages of this book. Foods that will enable you to improve the flavor of your cooking . . . foods that will provide new taste sensations . . . foods that will enrich your life.

    Happy shopping!

CATALOG OF FOOD

KITCHEN UTENSILS

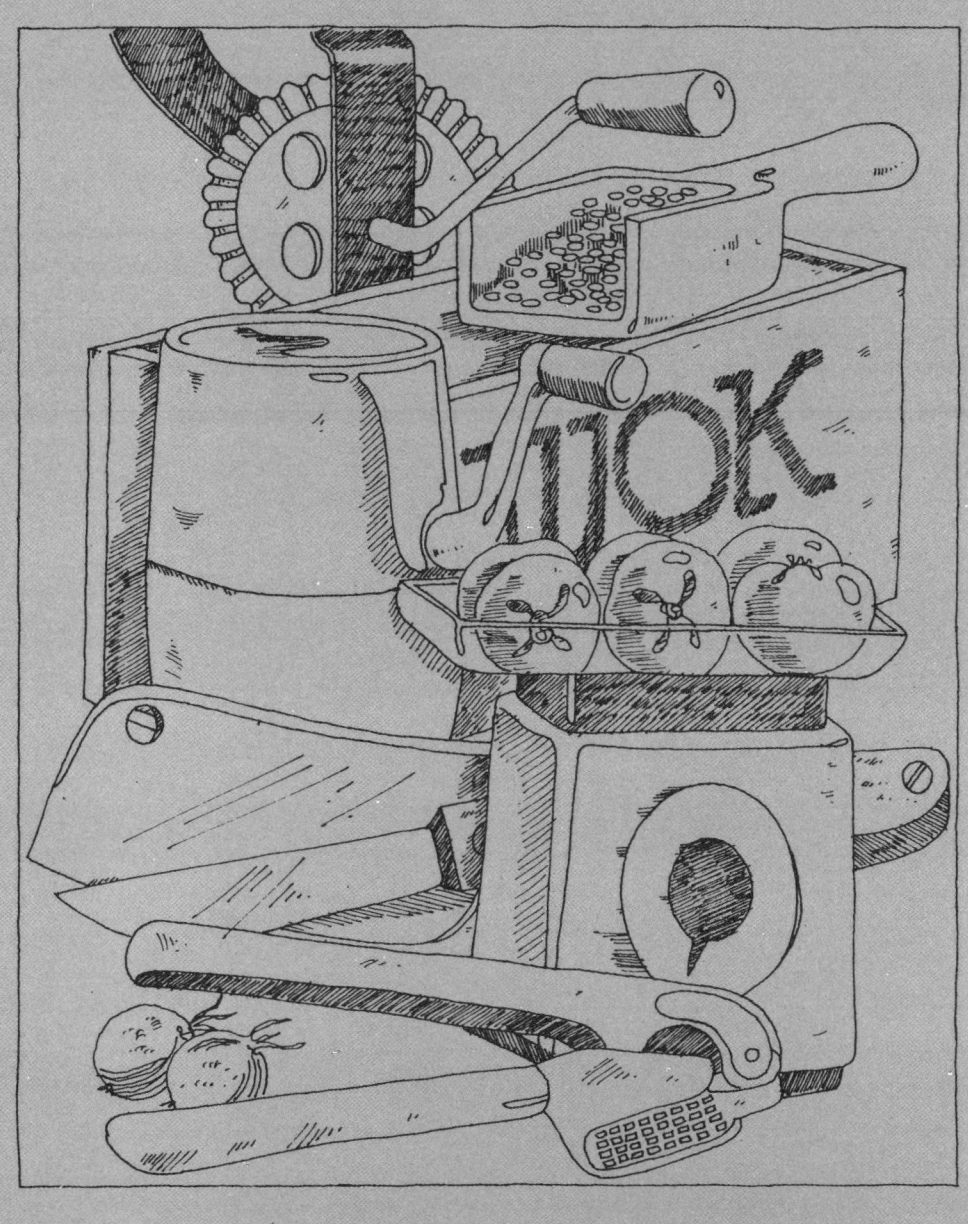

What's a kitchen without a wok? A madeleine pan? A goosefeather pastry brush? A spätzle machine? Or a magic mixer? The gourmet chef cannot be limited by the usual collection of cooking utensils for that person must be prepared for any recipe. And just in case you think that everything you need is available at the local five-and-dime store, take a look at the unusual items to be found in this chapter. Even at specialty stores, a good selection of kitchen utensils is difficult to find. The best way is to shop by mail, and mail-order companies have accumulated an outstanding variety of utensils and appliances for every need.

So, whether you're a gourmet cook looking for a duck press, or a beginner just starting to build your supply of kitchen utensils, send for some of the catalogs described below. (They have already done most of the work for you by scouring the manufacturers and importers for those unique or hard-to-find items that are so necessary to a well-stocked kitchen.) You'll have fun browsing and shopping in this world of fabulous utensils and gadgets, and you are sure to learn a thing or two!

CHRIS-CRAFT
Algonac, Mich. 48001

ICE MACHINE

"Joe the Bar Man" helps anyone to be a perfect host. Sign him on to your next party. "Joe the Bar Man" is a multipurpose appliance for 101 bar and kitchen uses. Actually three units in one, this versatile Sunbeam appliance mixes drinks, crushes ice for every purpose, and even feeds ice into the mixing jar while drinks are being prepared . . . and whips up a wide variety of foods—like salad dressings, appetizer dips, sauces, desserts, and many more. Mixing jar holds up to 48 ounces (3 pints) marked off in convenient 6-ounce graduations. Among the other admirable qualities "Joe" brings to the festivities are a folding ice chute door, jar cover with built-in strainer, molded pouring spout, and automatic dispenser pump that will pump out just the right amount of mixed drinks. Comes with 24-page instruction and recipe booklet. 120 volt, AC-DC. $34.95.

COUNTRY GOURMET
512 South Fulton Avenue
Mount Vernon, N.Y. 10550

THE GOURMET WOK

Enjoy the wok way of life—budget-wise, calorie-low, healthful! Chinese gourmet treats are hardy one-dish family meals! Sauté in mere drops of oil—see how a wok "gentles" foods, fluffs omelets, adds new flavor to old recipes! This 12″ stainless-steel pan has stay-cool jet handles. The aluminum lid has a matching knob, deep dome for perfect slow simmering. A hefty steel ring stand holds wok steady and assures even heat. $7.98.

WAKE TO FRESHLY BREWED COFFEE

This brainy thing goes to work while you are still sleeping! Simply set it for your breakfast hour, plug in your percolator, and go blissfully to bed. Has automatic shut-off and clear, easy-to-set dial. Easy instructions are included. Plugs directly into any ordinary outlet, with no wires to tangle. The tough plastic case is 3½″ ×2½″×1½″. U.L. approved, U.S. made by famed AMF—and they guarantee it. $6.98.

MORTAR AND PESTLE SET

Here they are, three gleamy white ceramic mortar and pestle sets. They're treasured by chefs for many gourmet chores and they give a smart new apothecary shop decor to any counter or shelf. They're terrific for easy blending of seasoned butter or gourmet paste, and for crushing and mashing garlic, herbs, and spices. Set of three, 1¾″, 3⅝″, and 4¾″ diameter bowls—$3.98.

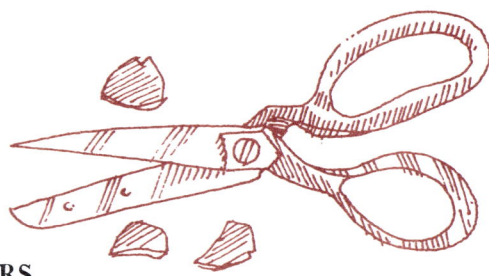

KRETZER-STRYER
104–20 Queens Boulevard
Forest Hills, N.Y. 11375

GIANT KITCHEN SHEARS

Most cooks have a pair of kitchen shears as a part of their everyday working tools. But most of the time, they don't really do the kind of job they need to. When you have a big job, where you want the blades to really sink into the roast, then here is a specially designed pair of shears with a long, flowing arc which will do the job. And to make matters even easier, one blade has a serrated edge so that you can easily cut through bones. The handles are of plastic and the blades are of chrome steel. This work horse is a real must for your kitchen. $18.00.

KITCHEN UTENSILS

SWISSMART, INC.
444 Madison Avenue
New York, N.Y. 10022

SPECIAL CHEESE CUTTER

This is an exquisite knife with two handles, one at each end. The blade, of stainless steel, is designed to cut wedges from wheels of firm cheese. It is beautifully made, and is 19″ long with handles made from rosewood. $24.00.

H. ROTH & SON
1577 First Avenue
New York, N.Y. 10028

FOOD CUTTERS

Have you ever been served gelée with a little block of aspic seemed to be specially designed? Well, perhaps they are. With these nifty little cutters, the next time you serve a gelée, you can cut the little pieces of aspic into alphabet-size pieces. These tin cutters range in size from 1¼″ to 1½″ in diameter and each is ¾″ deep. You can use them not only for gelée, but for truffles, pastry designs, cake designs, and anything else that moves you. They come in a neat little tin and steel container for easy storage. $39.50.

U. S. HEALTH CLUB, INC.
Yonkers, N.Y. 10701

MAKE YOUR OWN FRESH NATURAL PEANUT BUTTER!

Here is a kitchen appliance from Salton guaranteed to delight your family and provide excellent nutrition.

The leading brands of peanut butter sold in supermarkets today contain a number of additives that are completely unnecessary when you make your own, fresh, at home. There are no saturated fats or preservatives, no salt, no sweeteners or emulsifiers . . . and none are needed. You make the world's most delicious peanut butter just by pouring peanuts into the hopper, setting the dial to "chunky" or "smooth," and turning on the switch. Instantly the thick creamy rich peanut butter flows out into the receptacle, which has a cover for storing in the refrigerator. Fresh-ground peanut butter is one of nature's best and cheapest sources of protein. It is almost 28 per cent protein, low in saturated fats, and a good source of niacin, phosphorus, and magnesium. Your family and guests will love grinding their own nutritious, delicious, warm, crunchy or smooth peanut butter. Operates on regular AC 120 volt current, using only 130 watts. $29.95.

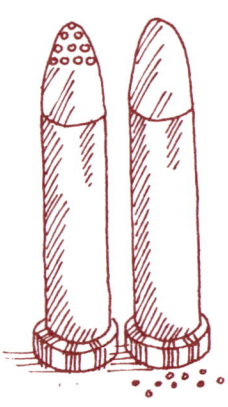

VERITY SOUTHALL
6041 Variel Avenue
Woodland Hills, Calif. 91364

OPEN FIRE WITH THIS PEPPER MILL AND SALT SHAKER

Every family should have at least one good pair of spice dispensers which really do the job. Here is an elegant pair of wooden cylinders which look very much like the cartridges used for guns. They come in enameled red, blue, or green with silver-colored tops. One is a salt shaker and the other is a pepper mill. Each one is 4¾" high and they cost $17.50 for the set.

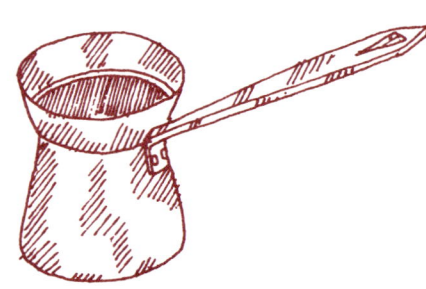

NORTHWESTERN COFFEE MILLS
217 North Broadway
Milwaukee, Wis. 53202

A TURKISH COFFEE MAKER

This small pot is frequently referred to as a Turkish coffee maker. It is also the traditional coffee maker for Greeks, Serbians, Arabians, and other nationalities. The brewing pot is usually made of brass or copper. The coffee, most often of a French-style dark roast, is combined in the pot with sugar and water. The brew is brought to a boil and removed from heat until the froth subsides. This procedure is repeated twice more, with the brewed coffee then being poured frothing into tiny demitasse cups for serving. Turkish Coffee Maker, Brass Ibrik: 5 ounces (3 cups)—$8.25. 8 ounces (4 cups)—$9.25. 12 ounces (5 cups)—$12.00.

See Beverage chapter for company description.

KITCHEN UTENSILS

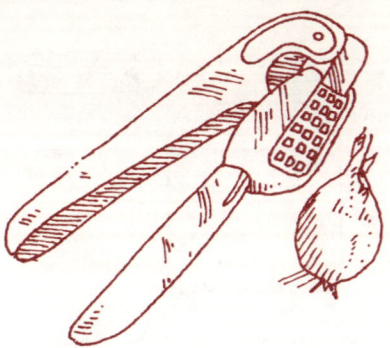

BECKER & BECKER
50 State Street West
Westport, Conn. 06880

GARLIC PRESS

Made from cast aluminum, with a curved, triangular container, this German-made garlic press will hold several cloves of garlic. What's even better is that you don't have to peel the cloves before pressing. And the holes of the press are large enough for easy cleaning. One other great benefit that this garlic press has is that it can also be used for onions. $5.00.

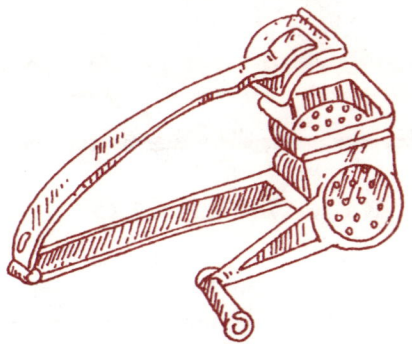

NICHOLS GARDEN NURSERY
1190 North Pacific Highway
Albany, Oreg. 97321

MOULI ROTARY MILL

It strains, mashes, pulps, rices! It's used with cooked vegetables, fruits, and nuts. An appliance that is a gourmet's delight. The only rotary mill on the market that has three easily replaceable mesh sizes with perforated cones for use on a wide variety of foods. The rotary mill snaps apart instantly for easy cleaning and changing of the perforated cone. It has a huge 2-quart capacity. And it's made of high-grade steel and acid-resistant coating. $7.50 postpaid.

K-TEL ELECTRONIC YOGHURT INCUBATOR

Now you can make your own delicious yoghurt, free of commercial thickeners and additives. All you need is powdered or whole milk, a couple of tablespoons of yoghurt starter, and presto, you have yoghurt. With the K-Tel, there is no risk of failure. It has special electric heating which keeps it at just the right temperature to make 1 full quart of delicious yoghurt. K-Tel is made of heavy-duty, break-resistant plastic that will give you years of service. Comes complete with electrothermal container and a jar that seals hermetically. Has an off-on light. $11.95 postpaid.

See Cheese chapter for company description.

ROWOCO
111 Calvert Street
Harrison, N.Y. 10528

BUTTER CURLER

If you've ever admired a hostess who serves sculptured butter at her dinner parties and wondered how she did it, then guess no longer. Here is a real butter curler, imported from Germany. It's only 7″ long with a curling loop designed to scrape across the top of hard butter to pull off a curl. It's only $2.00.

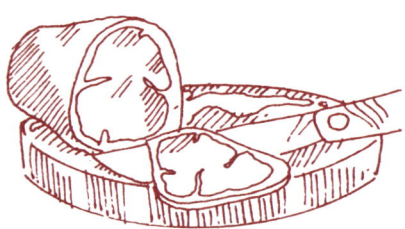

HARRINGTON'S
Richmond, Vt. 05477

THE NEW JUICE CATCHER CARVER

The flat surface of this handsome oval board slopes to the rear well where all the juices collect, out of the carver's way. The generous 18″×24″ size provides over 275 square inches of cutting surface for ample carving room with ample left over to stack slices for serving! Also available in a 16″×21″ size. The early Vermont finish displays both meat and fowl to their most appetizing advantage.

    16″×21″ .... $28.00 prepaid
    18″×24″ .... $29.35 prepaid

See Meat, Fish and Poultry chapter for company description.

GREAT VALLEY MILLS
Quakertown, Pa. 18951

CLAY COOKING POT

Now an ancient method gives you a new way of cooking. The Romertopf pot enables you to cook in clay—the natural way. This handsome, historic-looking pot takes the worry—and the difficulty—out of those hard-to-fix dishes. Turkeys,

KITCHEN UTENSILS 9

hams, roasts, all cook in their natural juices—with no need for extra fat. You'll save valuable vitamins and nutrients as well as calories. Weight-watchers will especially enjoy this cooking aid!

The Romertopf clay pot comes in two sizes and is accompanied by its own recipe book.

    small (3–4 pounds) $14.95
    medium (5 pounds) $19.95

For free catalog describing their many lovely food items, write the above address.

See Special Foods chapter for company description.

R-JAY PRODUCTS, INTERNATIONAL
7824 Thornton Road
P.O. Box 8471
Stockton, Calif. 95208

FRUIT PITTER!

Have you ever gotten tired and bored from pitting what seems to be an endless amount of pit-bearing fruit? Now you can make pitting easy and save yourself hours of work. Pit-o-Matic—the unique new pitter imported from Switzerland—is a convenient kitchen device designed to save you time and aggravation. Use it for cherries, dates, prunes, plums, olives, and other fruit. Cost of this amazing machine is only $18.95 plus $1.00 for postage and handling.

BISSINGER'S
205 West Fourth Street
Cincinnati, Ohio 45202

CHOCOLATE MILL

Add a new twist to your desserts with this chocolate mill. It works like a pepper grinder to shave flakes and curls of chocolate. In the kitchen you can decorate fancy cream pies, cakes, tarts, and pastry. And chocolate shavings melt faster in your double boiler than solid squares or even bits. On the dinner table the chocolate mill adds that personal finishing touch to parfaits, sundaes, whipped-cream and ice-cream desserts. Each mill comes packed with 12 ounces of Bissinger's Special Formula Chocolate, enough to last for many meals. A useful addition to your own kitchen, and a unique gift for your do-it-yourself gourmet friends. $7.50 plus postage and handling.

For ordering information and a complete catalog, write the above address.

LEKVAR BY THE BARREL
H. Roth & Son
1577 First Avenue
New York, N.Y. 10028

COOKIE CUTTER SETS

Now—for your cookie-baking fun—here's an invaluable aid: Lekvar's cookie cutter sets. These exquisitely designed and crafted sets are a must for anyone who loves to bake. With the Zoo Set, for instance, you'll have the pleasure of seeing your cookies "come alive" in the form of little elephants, ducks, camels, and nine other delightful creatures. Each piece is 1½" to 2½" in diameter, and the entire twelve-piece set costs just $4.95.

Lekvar also offers you over fifteen other varieties of cookie cutter sets, including special sets for Christmas and Easter. For a complete catalog, write the above address.

EVA HOUSEWARES, INC.
P.O. Box 2087
San Rafael, Calif. 94902

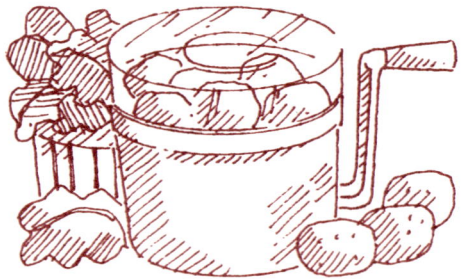

POTATO PEELER WITH LETTUCE SPINNER

Do two jobs for the price of one! The Eva potato peeler and lettuce spinner is a unique appliance which peels up to 2 pounds of potatoes in less than 3 minutes—and then, with a quick change of accessories, dries your lettuce instantly. Take your choice of red, orange, yellow, white, avocado, or blue—all designed to complement your kitchen decor. $25.00.

EVA HOUSEWARES, INC.
P.O. Box 2087
San Rafael, Calif. 94902

PARSLEY CHOPPER

The meals you serve can become more nutritious and appealing with the new Eva parsley chopper. This tastefully designed appliance can be used in the kitchen, but it's a welcome addition to any dining table. (Use it like a pepper grinder.) You and your guests will appreciate the convenience of fresh parsley—and you'll also have a fascinating conversation piece everyone will have fun using. Available in rich colors that will enhance any dining decor. Your parsley chopper is also useful for chervil, cabbage, mint, lemon peel, orange peel, baby food, eggs, Parmesan cheese, almonds, chocolate bars, and many other foods. $5.00.

NICHOLS GARDEN NURSERY
1190 North Pacific Highway
Albany, Oreg. 97321

SAFTBORN

An ideal appliance for all those who would like to extract juice from their surplus fruits and vegetables. Each Saftborn has a 12-pound fruit capacity. This remarkable new dejuicer extracts juice from all kinds of fruits, through a steam process. The juice comes out sterilized and ready to be canned or used for wine making. The juice is concentrated, clear, and free of impurities. Saftborn extracts juice from all hard fruits like apples, in about fifty minutes; soft fruits like berries, in twenty minutes. Color is preserved, and up to 90 per cent of the natural, health-giving fruit sugars are saved. The Saftborn can be used for steam-cooking vegetables, like cabbage and corn. They come out firm but fully cooked, with their flavor locked in and no loss of color. Delicious! Saftborns are sturdily constructed of hard-spun aluminum to prevent pitting, and will give many years of trouble-free service. They are guaranteed to satisfy. Postpaid prices: each Saftborn $43.50 west of Rockies; $45.50 east of Rockies.

See Cheese chapter for company description.

WILLIAM PENN GIFTS
6515 Castor Avenue
Philadelphia, Pa. 19149

AUTOMATIC ELECTRIC FONDUE SET

Ten-setting heat selections knob for no-guess cooking of your favorite fondue. "Burner-with-a-brain" heat sensor automatically holds exact for dessert fondues, cheese fondues, sauces, and meat fondues. Porcelain-on-aluminum pot is lined with no-stick, no-scour Teflon II and has a removable antisplatter cover notched for forks. Complete with recipe book, 4 color-coded forks, a 14-ounce package of cheese fondue, and 3 bars of Tobler fondue chocolate—all ribboned and bowed in the famous William Penn manner. $39.95 delivered.

ABOUT THIS COMPANY

Gifts of fine foods from William Penn are a delightful, delicious, and welcome surprise to any recipient, and no gifts are easier to select and send. Choose from the widest variety anywhere for each person on your list. Then be assured that each gift is distinctively different because it's an exclusive William Penn creation —sold only in William Penn shops and through their national mail-order service.

PAPRIKAS WEISS IMPORTER
1546 Second Avenue
New York, N.Y. 10028

THE DUMPLING MACHINE

In Eastern Europe, the way of a woman with dumplings is as important as her charm or beauty! Considerable competition surrounds the making of one variety, spätzle, the tiny, tender morsels of dumpling dough that are tossed with butter, cheese, and sour cream to grace a platter of chicken paprika or goulash. Just as the quality varies from kitchen to kitchen, the size of these dumplings changes from country to country. Hungarian cooks make the biggest spätzles; Germans, the smallest. Imported Hungarian-style spätzle machine—$11.98. Imported German-style spätzle machine—$11.98.

See International Groceries chapter for company description.

KITCHEN UTENSILS

PAPRIKAS WEISS IMPORTER
1546 Second Avenue
New York, N.Y. 10028

HACHE-TOUT

Are you tired of choppers that don't keep their promises—dull blades or poor chopping ability? Well, here's a chopper that really works. Its five stainless-steel blades are razor-sharp and are meant for doing all of those hard-to-do jobs such as slicing, dicing, mincing, etc. It's got a dishwasher-proof handle, so pop it in the dishwasher when you're finished. It's $3.98.

See International Groceries chapter for company description.

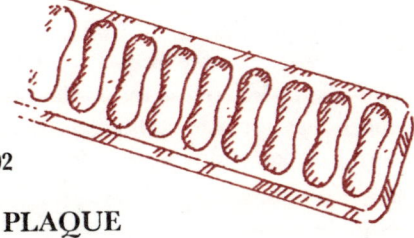

WILLIAMS-SONOMA
Mail Order Department
532 Sutter Street
San Francisco, Calif. 94102

THE MADELEINE PLAQUE

This heavy French tin is named for the famous nineteenth-century cook who created the recipe for this small, rich cupcake that Proust made famous. The twelve impressions are especially designed to hold these 3″ cakes. And in case you wonder how to make these tasty treats, a recipe is included. $5.75.

See company description in this chapter.

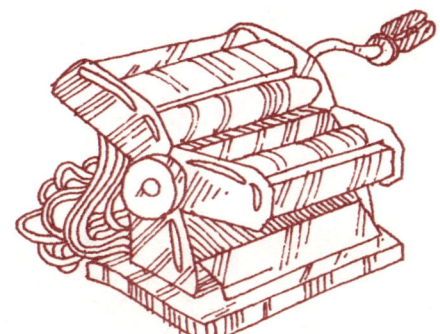

PAPRIKAS WEISS IMPORTER
1546 Second Avenue
New York, N.Y. 10028

HOME NOODLE-CUTTING MACHINE

Available exclusively from Paprikas Weiss, there is no other machine in the world designed for home use that will cut noodle dough into such slender strands. This precision-built machine is so finely geared that only a hairline space is provided between blades and a holder. The result is ultra-thin noodles or vermicelli, so lip-smacking good in soup, produced in your own kitchen in only a few minutes. $59.95.

See International Groceries chapter for company description.

14   KITCHEN UTENSILS

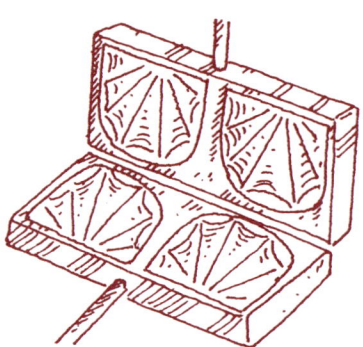

WILLIAMS-SONOMA
Mail Order Department
532 Sutter Street
San Francisco, Calif. 94102

WAFFLE IRON

Show your love at the breakfast table with a heart-shaped waffle iron from Denmark. This delightful appliance will bring you many hours of fun and fine eating. Made of cast aluminum with a no-stick Teflon lining, the waffle iron is easy to use. Serve your waffles with melted butter and syrup on a cold winter morning . . . with strawberries and whipped cream for a special Valentine's Day treat . . . or with ice cream and honey for a refreshing summer dessert. Any way you fix them, they'll always turn out just right. Perfect for parties and buffets. Cost is just $17.00 plus $1.75 for shipping.

ABOUT THIS COMPANY

Williams-Sonoma has been dispensing culinary inspiration for nearly twenty years at their San Francisco address. In addition, they have two other locations, Beverly Hills and Palo Alto. They carry a fascinating array of cooking and serving equipment from Germany, Italy, England, China, and North Africa. Write for their free catalog.

BREMEN HOUSE, INC.
200 East 86th Street
New York, N.Y. 10028

UNIVERSAL SLICING MACHINE

Now you can slice almost any food in seconds—quickly and easily—with the amazing Universal Slicing Machine. This remarkable appliance is the perfect way to cut cheese, ham, sausage, bread, vegetables—in fact almost any food you like. It slices thin or thick—to your taste—and can be compactly folded when not in use. And your Universal Slicing Machine costs only $60.

See Condiments, Spices, and Syrups chapter for company description.

BREMEN HOUSE, INC.
200 East 86th Street
New York, N.Y. 10028

SPÄTZLE PRESSE

If you like to cook from scratch, you'll appreciate a Spätzle Presse. This handy appliance lets you make spätzle, noodles, mashed potatoes, juices, and many other foods quickly and easily. Your Spätzle Presse comes with complete instructions and costs only $14.95.

See Condiments, Spices, and Syrups chapter for company description.

COLONIAL GARDEN KITCHENS
270 West Merrick Road
Valley Stream, N.Y. 11582

OLD-FASHIONED IRON POTS

Cast-iron cookware . . . the age-old secret of fine cooks everywhere, and as American and durable as the Rockies. Provides even heat for better cooking, enhances food flavors, too. Two-quart Dutch oven is the famous Early American "Porridge pot" that made Great-Granny's cooking something to be acclaimed. Cooks evenly with little or no water (preserving precious nutrients). Turns out succulent pot roasts, stews, casseroles, and hearty meals your family will thrive on. Delight your family and guests with the tenderest, juiciest chicken dishes ever —prepared in the 10" chicken fryer. All these cast-iron pieces are extra-heavy weight, preseasoned, and come with domed, heat-resistant glass covers. The fascinating *Cast Iron Cookbook* is a treasury of lore and recipes from Nantucket Chicken Chowder to "The West's Best Chili."

2-quart Covered Dutch Oven	$6.95
4-quart Covered Dutch Oven	$7.95
10" Covered Chicken Fryer	$6.98
Cast Iron Cookbook	$3.95

ABOUT THIS COMPANY

Colonial Gardens has been in business for many years, supplying indispensable utensils to help you master every cooking and serving challenge. Their guarantee states: "Any item you select must please you 100 per cent or I do not want you to keep it! Return it for immediate refund—no questions asked. (We are members of the New England Mail Order Association, and are Dun & Bradstreet rated AAA-1.) Bank reference: Northern Bank of North America."

KITCHEN UTENSILS

PAPRIKAS WEISS IMPORTER
1546 Second Avenue
New York, N.Y. 10028

THE MAGIC MIXER

Here's a high-powered appliance that can turn the novice into an experienced cook, and save the expert cook hours of preparation time. This remarkable machine can do just about anything, yet it requires only a small amount of counter space.

Its two mechanisms accommodate attachments for blending, liquefying, mixing, beating, peeling, and cutting. But wait, don't put it away! It also makes ice cream, presses pasta, shapes pastries, fills sausages, grates cheese, purées, and acts as a sieve. Then, snap on a six-cup plastic jar, and grind coffee or extract juice. The limitless number of kitchen feats that this device deftly performs makes it worthy of its "magic" name. Nearly all the attachments are dishwasher safe. $250.

See International Groceries chapter for company description.

COLONIAL GARDEN KITCHENS
270 West Merrick Road
Valley Stream, N.Y. 11582

SAN FRANCISCO HERB JARS

Keep your herbs in the dark to ensure lasting freshness of taste, color, and bouquet. Six handcrafted stoneware jars with wide-mouth airtight cork tops protect herbs from moisture and light. Natural clay color with rich brown glaze trim lends an artistic "California touch" to your kitchen. Titles read "Cloves," "All Spice," "Cinnamon," "Ginger," "Nutmeg," "Paprika." 3¾" high. Set of six—$7.98.

See company description in this chapter.

KITCHEN UTENSILS

COLONIAL GARDEN KITCHENS
270 West Merrick Road
Valley Stream, N.Y. 11582

CONTINENTAL TEA SERVICE

Tea was meant to be appreciated for its jewel-like amber color as well as for its distinguished flavor. Tea is beautiful, and it took our European friends to show us how to add this new dimension to the art of tea drinking. So elegant and utterly simple in design that this teapot is in the Museum of Modern Art collection. Made of heat-resistant Jena glass from Germany's most honored glassworks. Graceful clear-glass teapot holds 1 quart, has its own infuser to keep tea clear. Lovely clear tea mugs have easy-grasp handles, attractively etched design.

> Teapot $18.95
> Tea mugs, set of 6 .... $10.95
> Tea saucers, set of 6 .... $ 8.95

See company description in this chapter.

JAMES G. GILL COMPANY, INC.
204–210 West 22nd Street
Norfolk, Va. 23517

AROMATIC COFFEE MILL

If you like your coffee fresh-ground, you'll want this new Braun electric coffee mill. It's precision-designed, with a control dial that lets you select any one of nine different settings for the right grind. Whether your coffeepot is filter, drip, or percolator, the Braun aromatic coffee mill will unlock the full, rich flavor inside each coffee bean. With this miraculous machine, you'll enjoy a fresh, hearty cup of coffee every time. The Braun aromatic coffee mill costs only $35.00 and has been approved by the Pan American Coffee Bureau.

> *See Beverage chapter for company description.*

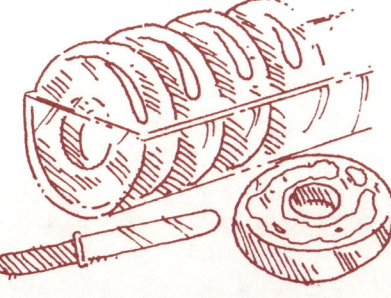

COLONIAL GARDEN KITCHENS
270 West Merrick Road
Valley Stream, N.Y. 11582

BAGEL CRADLE

Here's an attractive, convenient way to serve your bagels: the Bagel Cradle. Made of clear, sparkling lucite, this handy kitchen aid is a decorative addition to your dining table. Use it for bagels . . . sweet rolls . . . English muffins . . . doughnuts . . . biscuits . . . even chocolate chip cookies. Whatever you put in it, the Bagel Cradle is sure to give you many years of use—at breakfast time, snack time, or anytime. The Bagel Cradle measures 10″×4″×3½″ and holds six bagels. Cost is $5.95.

See company description in this chapter.

COLONIAL GARDEN KITCHENS
270 West Merrick Road
Valley Stream, N.Y. 11582

ELECTRIC HOME GRAIN MILL

Mmmm . . . when's the last time your family enjoyed home-baked bread with home-ground flour! You can bring a delightful variety of farm-fresh, fluffy ground meals and flours to your daily cuisine with no effort at all. Grind rice, corn, millet, wheat, sesame, sunflower—many, many more seeds and grains. Push-button control gives a steady flow and takes only *1 minute per pound.* Choose fine or coarse grind . . . add zestful nutrition to your menus! Safety switch, 140 watt AC motor, U.L. approved. 1-year guarantee. $29.95.

See company description in this chapter.

WILLIAMS-SONOMA
Mail Order Department
532 Sutter Street
San Francisco, Calif. 94102

"SMOKING" AT HOME

It's easy, it's inexpensive, and the results are delicious. All you need is an *electric smoker,* brine, wood chips, and something to smoke.

An outdoor location for the smoker such as a patio or porch is ideal, or just set it by the back door. Thin salmon fillets take about four hours to cure, thick ones about eight or ten hours.

You can suspend a whole turkey in the smoker by removing the racks—it will smoke up to twenty pounds of meat at a time—or make jerky if you have an elk handy. (James Beard has smoked really sensational hickory-flavored lean bacon in it.) You can also add a fine flavor to roasts or ribs by smoking them for an hour before cooking, and even wieners benefit from this treatment.

KITCHEN UTENSILS

The smoker is shipped completely assembled and ready to use. It has an all-aluminum exterior and an all-steel interior. Complete instructions are included. together with four 8-ounce sacks of smoker chips.

Smoker measures 24″ high by 12″ square and includes outer case, lid, three grills, drip pan, heat element, electric cord, and pan for fuel. $38.50.

Smoker Chips are available separately. Hickory (2 pounds), Apple (2 pounds), Alder (1¾ pounds), or Cherry (2 pounds). Any three bags, your choice, please specify. $6.00.

See company description in this chapter.

NORTHWESTERN COFFEE MILLS
217 North Broadway
Milwaukee, Wis. 53202

TEA INFUSER

The market place is glutted with a bewildering array of tea brewing aids. Since there is nothing complicated about them, most work, but some work beter than others. Northwestern strongly recommends stainless steel over the chromed brass and aluminum.

 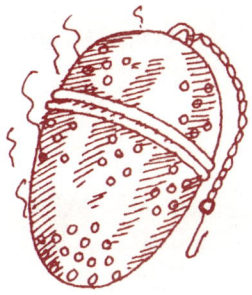

Spoon The stainless-steel spoon has a side hinge with a firm-closing clasp. The German-made infuser is lightweight and will brew several cups of tea in succession if the spirit moves. $2.75.

Ball This stainless-steel tea ball has a unique hinged closure making it easier to fill and empty. Its chain is securely fastened and it should last for a reasonable time to forever. 1¾″ round, 1″ depth. $2.00.

See Beverage chapter for company description.

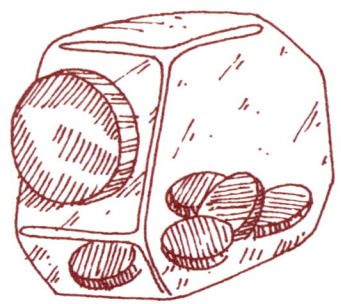

MOTHER'S GENERAL STORE
P.O. Box 506
Flat Rock, N.C. 28731

PENNY CANDY JAR

If you've ever seen pictures of an old country store, you know candy jars were an indispensable part of the decor. Now you can have your own penny candy jar—created in the same style. Made of clear glass with a metal lid and red knob, this handy container is a useful and decorative kitchen appliance. Use it for candy, cookies, doughnuts, muffins, brownies, carrot and celery sticks—even as a terrarium. Whatever you put in it, you'll find it's an attractive addition to your kitchen. The penny candy jar weighs 7 pounds and costs $4.35.

See Food Kits chapter for company description.

NORTHWESTERN COFFEE MILLS
217 North Broadway
Milwaukee, Wis. 53202

TWO-PART COFFEE MAKER

This Italian-made coffee maker is of two parts, one for holding the water and one for condensing the steam into brewed coffee. This particular brand is of cast aluminum with brass pressure relief vents and heavy-duty rubber gaskets. The sides are faceted for easy separation of the sections. Water placed in the lower compartment will turn into steam as it is boiled under pressure. The steam is vented through finely ground coffee held above the lower compartment and then condensed into rich, strong espresso coffee in the upper section.

Espresso Maker: 11 ounces (6 cups)	$ 6.75
Fine Grind: 18 ounces (9 cups)	$11.50
29 ounces (12 cups)	$16.75
Replacement Gaskets, any size	$ .50

See Beverage chapter for company description.

KITCHEN UTENSILS

NORTHWESTERN COFFEE MILLS
217 North Broadway
Milwaukee, Wis. 53202

TEA KETTLE

An essential for those with Melitta-type filter coffee makers and for those who brew tea the proper way (one pot for boiling the water, one pot for brewing the tea). The water-boiling "tea kettle" should be of durable top-quality stainless steel, of a convenient size, easy to clean, and well balanced to pour easily. $14.75.

See Beverage chapter for company description.

WILLIAMS-SONOMA
Mail Order Department
532 Sutter Street
San Francisco, Calif. 94102

FRENCH BREAD PANS

Quilted aluminum French bread pans prevent long loaves from spreading into each other in the oven and help produce a fine, crunchy crust. The recipe that comes with the pans advises using the pans as forms in which to raise the bread as well as for baking. When the loaves have risen, pull the pans out from under the cloth and flip the loaves over, back onto the pans. In this way, the soft underside, which has not been exposed to the air, will be uppermost and the bread will consequently rise a great deal higher in the oven.

Set of two double French bread pans (makes 4 loaves) 18″ long. Please measure your oven before ordering! $7.95.

Canvas pastry cloth (20″×24″) and French steel pastry scraper with wooden handle. $6.00 for the pair.

See company description in this chapter.

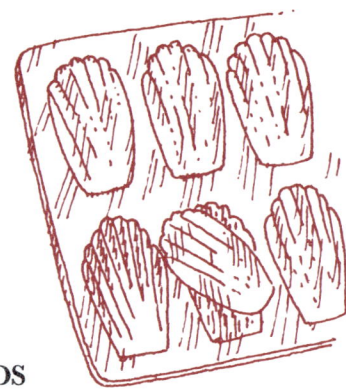

WILLIAMS-SONOMA
Mail Order Department
532 Sutter Street
San Francisco, Calif. 94102

PETITS FOURS MOLDS

Petits fours are a splendid way to finish an elegant dinner. They don't compete with the main course as a large and elaborate dessert might, and they are—or should be—as pretty to look at as they are delicious to eat. It is almost impossible to buy quality petits fours, but if you have the patience and a little imagination, you can make really beautiful selections at home, of the very finest ingredients.

The base is usually Genoese sponge cake, which doesn't crumble and keeps well, and the little cakes are filled with buttercream, coated with sieved apricot jam, and then covered in fondant icing or coating chocolate.

Julia Child gives very easy directions for making professional fondant icing. Williams-Sonoma definitely recommends using the couverture chocolate (see Confections chapter) which melts to the proper consistency.

Fifty assorted petits fours molds come in their own tin box, are approximately 1½″ in diameter, and can also be used for chocolate or marzipan confections. $13.50.

See company description in this chapter.

COLONIAL GARDEN KITCHENS
270 West Merrick Road
Valley Stream, N.Y. 11582

STAINLESS-STEEL KITCHEN CENTER

Like having a master chef in your kitchen! Whip up velvety sherbets from juicy fresh fruit and berries. Make raw vegetable salads, still tangy and fresh from the garden. Grind coffee, whole grains or crunchy cereals, or the finest cake flour . . . make baby foods or follow special diets with ease. All in just seconds! Your imagination is the only limit! No need, ever again, for juicers, grinders, food mills, ice choppers, or blenders. This amazing appliance replaces them all! The secret of this versatile kitchen center is its power and ability to reverse direction of the blades *instantly* . . . for the smoothest, fastest liquefying, shredding, grinding, chopping. Economical, healthful too—as all those precious nutrients that slip away in fruit and vegetable skins you throw out now blend deliciously for fuller flavor, more nutrition. Unique hammer-edge blades never need sharpening. Self-cleaning gleaming stainless steel, country-wide service, 5-year guarantee. Book of 800 exciting recipes included. $89.95.

See company description in this chapter.

OTHER SOURCES FOR KITCHEN UTENSILS

EKCO HOUSEWARES COMPANY
9234 West Belmont Avenue
Franklin Park, Ill. 60131

NORRIS INDUSTRIES
5119 District Boulevard
Los Angeles, Calif. 90040

LA CUISINIERE
867 Madison Avenue
New York, N.Y. 10021

HOFFRITZ
20 Cooper Square
New York, N.Y. 10003

BAZAAR DE LA CUISINE
1003 Second Avenue
New York, N.Y. 10022

BRIDGE KITCHEN WARE COMPANY
212 East 52nd Street
New York, N.Y. 10022

KITCHEN GLAMOUR
26770 Grand River Avenue
Detroit, Mich. 48240

GEORG JENSEN
601 Madison Avenue
New York, N.Y. 10010

B. ALTMAN & CO.
Fifth Avenue & 34th Street
New York, N.Y. 10016

SAKS FIFTH AVENUE
Fifth Avenue & 50th Street
New York, N.Y. 10022

BLOOMINGDALE'S
Lexington Avenue & 59th Street
New York, N.Y. 10022

HANDCRAFT FROM EUROPE
P.O. Box 372
Sausalito, Calif. 94963

GOURMET LIMITED
376 East St. Charles Street
Tombard, Ill. 60148

SHERRY-LEHMANN
679 Madison Avenue
New York, N.Y. 10021

ZABAR'S
2245 Broadway
New York, N.Y. 10024

SALTON
60 East 42nd Street
New York, N.Y. 10017

MEAT, FISH AND POULTRY

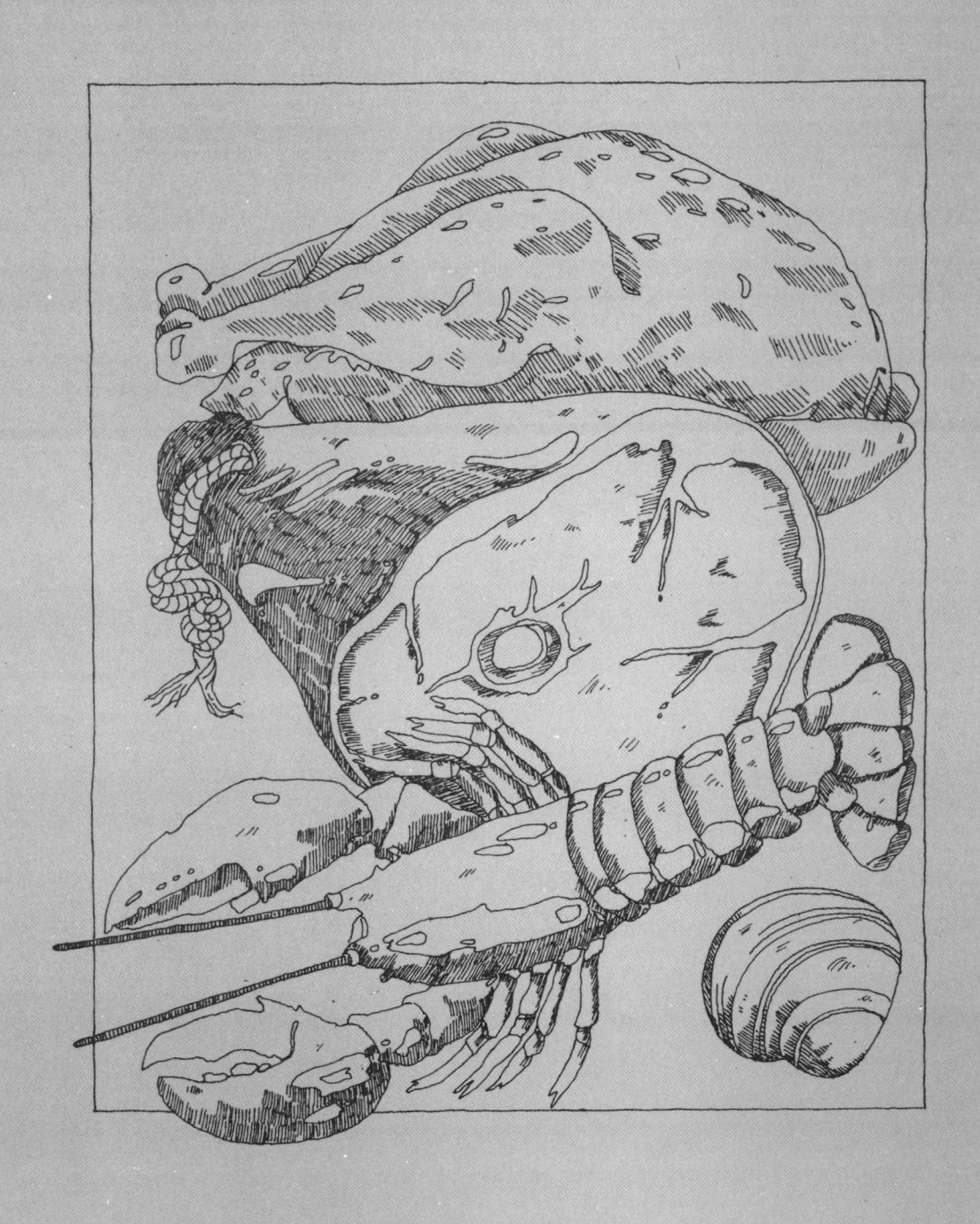

Smoked pheasant . . . Florida stone crab claws . . . ham . . . Cornish hen . . . Chateaubriand . . . lobster tails . . . are just a few of the many gourmet foods you can buy by mail. No longer are these delicacies reserved for those lucky enough to live in cities with a specialized gourmet food store. Now there are a multitude of gourmet mail-order food companies which specialize in providing you with the very finest available in meat, fish, and poultry. It seems that Americans have learned that there is more to eating than just plain steak and potatoes!

As man has learned more sophisticated techniques for breeding, feeding, catching, slaughtering, and preserving livestock, fish, and poultry, the quality of these foods has dramatically improved. Whoever thought you would see the day when fresh-caught lobster or organically raised livestock would be winging its way across the country to be part of the specially planned meal you are serving.

When it comes to gourmet foods, there are many companies which have been in business for generations and which are still family-owned. Omaha Steaks of Nebraska is one. But whether you order one of their naturally aged, beautifully marbled steaks, or a hickory-smoked ham from Smithfield, or a fantastic lobsterbake from Saltwater Farm, or a smoked duck from the Fin 'n Feather, you can't go wrong.

And what hostess wouldn't be absolutely thrilled to receive a brace of plump, white-skinned quail specially raised on domestic prairie lands or a half-dozen T-bones? These foods are also a wonderful way for you to make an ordinary meal into a gourmet feast. So if you're a connoisseur, or a lover of meat, fish, and poultry delicacies, then you won't want to miss out on this chapter.

OMAHA STEAKS INTERNATIONAL
4400 South 96th Street
Omaha, Nebr. 68127

GOURMET MEATS FROM THE BEEF CAPITAL OF THE WORLD

126 years in the meat business assures the quality of the steaks you order from the Simon family at O.S.I. 24 of those years spent serving the nation's quality-conscious meat-lovers by mail assures the safety and freshness of your delivered orders.

Prime Rib Roast To assure you utmost pleasure, only the first five ribs are included in this incomparable roast. Marbled, meaty, prime or choice corn-fed beef at its very best—with only enough fat left to enhance the flavor. It's oven-ready. (Easy cooking and serving instructions with order.) One 12–13-pound (minimum 12 pounds) prime rib roast—$59.50.

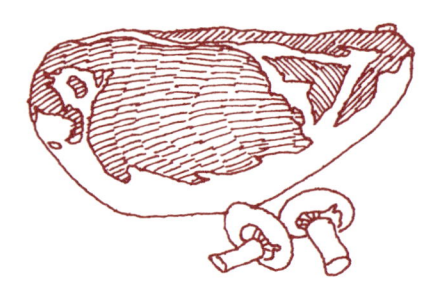

Boneless Strip Sirloins America's favorite cut. A specialty of the finest dining rooms. Naturally aged, beautifully marbled, and trimmed with just enough fat for exceptional flavor. These steaks are tender, juicy, and so tasty you'll be amazed. Twelve 7-ounce portions, ¾″ thick—$43.50.

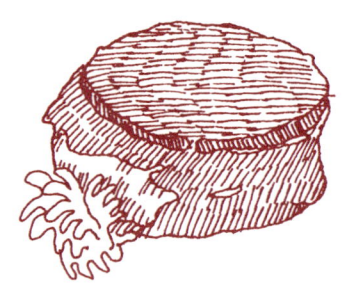

Filet Mignon Succulent center cuts from the choicest beef tenderloins aged naturally to perfection. So tender that using a knife is a social gesture rather than a necessity. Each order includes James Beard's recipe for béarnaise sauce. Sixteen 6-ounce portions, 1¼″ thick—$52.00.

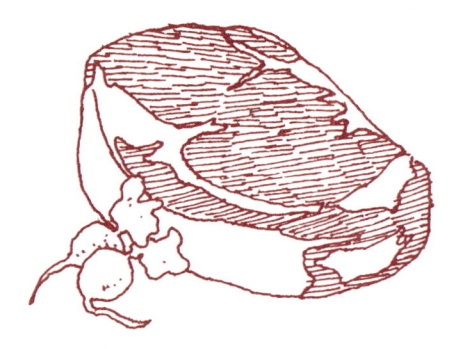

Fillets of Prime Rib (Rib Eye Steaks) Imagine a steak cut from the tenderhearted center of a luscious prime rib roast. Think of the delicate texture, the luxurious richness, the succulent flavor it would have. Why imagine it? Simply order a selection of fillets of prime rib and try one for yourself! You'll discover a slightly different steak that's equally good at a white tie affair or a casual cookout! Eight 12-ounce portions, 1¼″ thick—$52.00.

MEAT, FISH AND POULTRY

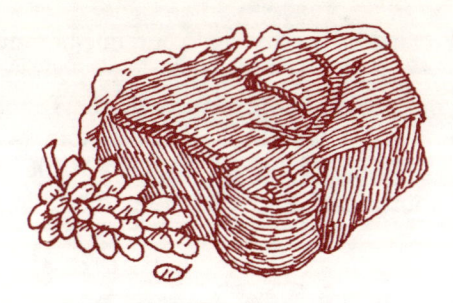

29

Top Sirloins These top sirloins are a legacy! And now, an improved cutting method developed by the Simon family makes them even better! The result is a more tender, more uniform steak. These are thick and chunky center cuts of the top sirloin, adorned with just a thin crown of fat. Steak lovers appreciate their incomparable corn-fed flavor and pleasing texture. Our natural aging makes the difference—imparting a mellow tenderness and enhancing the taste. Ten 10-ounce portions, 1¼″ thick—$48.50.

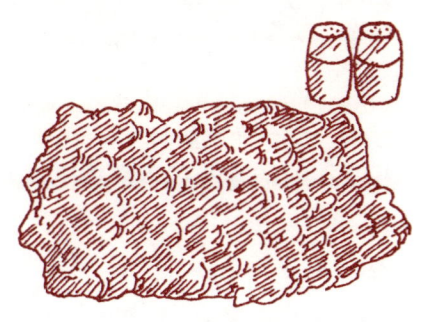

Ranch Minute Steaks They pamper these steaks like you wouldn't believe. They're aged, trimmed, then individually wrapped. Every beef lover deserves a supply—for that late snack, midday lunch, or hearty breakfast. Give them a good sizzling broil. Serve on bread, stuff them, cover with sauce, use your own touch. Twelve 5-ounce portions, ½″ thick—$25.50.

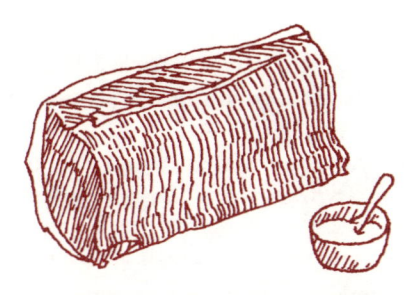

Chateaubriand This regal cut of meat originated in France's great wine-making regions. Boneless and oven-ready, it's the very center cut of the tenderloin, crowned with only enough fat to complement the subtle flavor of the meat. Serves four to six. (Easy cooking and serving instructions with order.) One 4–4½-pound (minimum 4 pounds) Chateaubriand—$44.00.

Veal Saddle Roast You'll be sure to win your guests' admiration when you serve this delicate boneless veal roast, expertly trimmed from the finest milk-fed calves. It's this special milk feeding (developed by the Dutch) which gives the veal its firm, lean texture and creamy-white color. (Note: We portion the lean veal "saddles" into 2½–3-pound roasts that vary a little in size and shape.) Recipes included with your order. One 2½–3-pound (minimum 2½ pounds) veal saddle roast—$31.50.

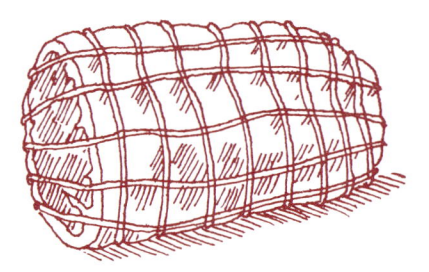

Boneless Leg of Spring Lamb In response to customer suggestions, one of our meat experts has developed an improved trim for this succulent delicacy. As always, those who appreciate the best in spring lamb will be pleased with this delightfully sweet and tender meat. Oven-ready. We have removed the bone and tied the leg for roasting. (A new recipe by James Beard enclosed with order.) One 6½–7½-pound (minimum 6½ pounds) leg of spring lamb, boneless and tied—$40.50.

Quail Originally Asian game birds, these plump, white-skinned quail are raised on domestic prairie lands. Their flavor is rich and savory—without being gamey. Serve these delights at your next dinner party for an exotic change of pace. Plan on a brace of quail per serving—two meaty, succulent birds per person. Completely cleaned and ready to cook. Easily prepared—in minutes. (Easy cooking and serving instructions with order.) 12 (4–6 ounces) quail—$43.50.

Irish Smoked Salmon Firm, smoked to perfection, and cured to an uncompromising exactness—Irish Smoked Salmon meets these supreme tests. Salmon from sparkling northern waters of Ireland assure you fresh flavor, easy-to-slice firmness. Only the finest of these are expertly selected for you and prepared with traditional artistry for curing and smoking. Delicious as hors d'oeuvres—or as a breakfast treat. Shipped October through March. Side of 2–2½ pounds (minimum 2 pounds) of smoked salmon, $29.50.

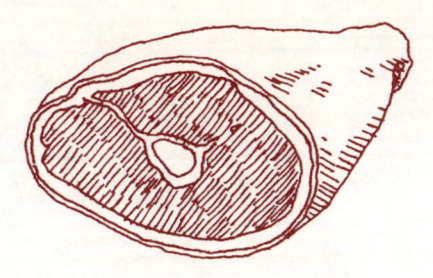

Bone-in Country Ham A large, mellow, tender ham. A wonderful feast for holidays, all occasions. Short-shanked, closely trimmed, naturally aged. It's cured to perfection, then hickory-smoked the old-fashioned way. This does require cooking, but it's worth it. Available all year round. 13–15 pounds—$39.50.

Country-style Slab Bacon This is the savory bacon most people only dream about. It's expertly selected for leanness and tenderness, then cured and hickory-smoked. You get a full 5 pounds in two 2½-pound vacuum-sealed packages. Sliced to the thickness you prefer. Available all year round. Two 2½-pound pieces—$15.50.

Completely Cooked Smoked Turkey Unusually tender birds, delicately prepared to bring out an elegant flavor. Slowly smoked to retain their natural juices while imparting old-fashioned goodness. Have a party or make an occasion with this delicacy and hope that you have some leftovers to enjoy! Shipped October through March. 9–10 pounds—$22.50.

Canadian-style Bacon Elevate the first meal of the day to an elegant occasion with this meaty Canadian-style bacon. It's one of the most perfect products of its kind. The trim is exceptional; the flavor, a delight. You can even slice and serve it as an hors d'oeuvre, since it's fully cooked and ready to eat. Shipped October through March. Two 2-pound pieces—$19.95.

Country-style Ham Steaks Tender, juicy center slices, dry-cured the southern country way. Each ham is hand-rubbed to control the salt . . . and a secret dry-curing formula guarantees these ham steaks are so tasty you'll want them often. For brunch; for a spur-of-the-moment dinner. Quick and easy to prepare. Available all year round. Six 6-ounce ham steaks—$17.50.

See Special Foods chapter for company description.

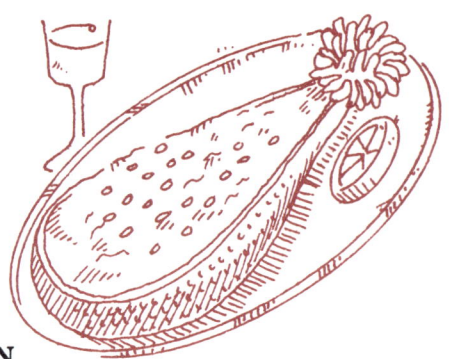

SALTWATER FARM
York Harbor, Maine 03911

IRISH SMOKED SALMON

If you flip over fine fish, you'll want to try this delicious smoked salmon. Imported from Ireland, this savory delicacy is a must for that elegant dinner table. It's so delicious, your guests will keep coming back for more. Each side of salmon weighs 2 to 2¼ pounds, and can be refrozen without damaging its high-quality taste or texture. One side of salmon costs $23.50; two sides salmon, $43.55; and ten or more sides, $21.15 per side. For airmail delivery west of the Mississippi and south of North Carolina, add $2.00

See company description in this chapter.

SALTWATER FARM
York Harbor, Maine 03911

CLAM STEAMER

Now you can cook your clams to perfection—with the Saltwater Farm clam steamer. This efficient appliance lets you bring out all the natural juices and flavors that make clams such a succulent, tasty dish. And it's uniquely constructed in two sections: one for the clams and the other for clam broth. If you love to cook clams, you'll love this handsome clam steamer. Available in 4-gallon size for $21.15 or 8-gallon size for $27.30.

See company description in this chapter.

SALTWATER FARM
York Harbor, Maine 03911

STUFFED RED SNAPPER

These ocean-fresh fish are the most prized of any for eating. Immediately packed in ice when caught, the snapper is then filled to overflowing with a delicious crab

MEAT, FISH AND POULTRY

meat stuffing and quick frozen to seal in all its flavor. Four one-pounders are ample for six to eight people. Fish are completely prepared and ready for baking or broiling. Recipe enclosed. Four 1-pound red snappers—$39.95.

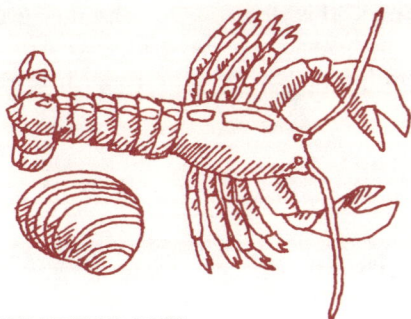

A FANTASTIC LOBSTERBAKE

One of the best lobsters in the world comes from Maine. Each one is 1⅛ pounds of the most succulent, tender, mouth-watering lobster you can imagine. Included is a half peck of butter-sweet Maine steamer clams. They are gently nestled in fresh, hand-picked rockweed and packed into a convenient steamer can. Whether it's for yourself or as gifts to friends, relatives, and business associates, a Saltwater Farm lobsterbake is an unforgettable treat. 8 lobsters, ½ peck clams—$48.70.

ABOUT THIS COMPANY

Saltwater Farm is located in Maine and has been specializing in gourmet seafoods for twenty-six years. Send for their free, full-color catalog.

SALTWATER FARM
York Harbor, Maine 03911

LOBSTER TAILS AND FILET MIGNON

Here's a delicious way to get two gourmet meals in one: with the Saltwater Farm lobster tails and filet mignon. These delicious 4-ounce lobster tails are from large, juicy Maine lobsters. They're cooked to perfection in Saltwater Farm kitchens to give you a hearty, savory flavor that can't be surpassed. And Saltwater Farm also brings you the tender, succulent beef of aged tenderloins in its prime-quality 6-ounce filet mignon. These superb steaks are specially selected for their hearty taste. Order six filets and six lobster tails for only $40.75 or ten filets and ten lobster tails for $66.75.

See company description in this chapter.

VERMONT COUNTRY STORE
Weston, Vt. 05161

SCOTCH-STYLE FINNAN HADDIE IN CREAM SAUCE

Back again after eight years, this delicious, toothsome peat-smoked haddock is prepared in rich cream sauce. Ready to heat and serve on toast, rice, or potatoes. A gourmet treat for a traditional Scottish or English breakfast. One 15-ounce tin—$1.50.

SMITHFIELD HAM AND PRODUCTS CO., INC.
Smithfield 8, Va. 23430

SMITHFIELD PANTRY SHELF PACKAGE

A beautifully boxed, large assortment of famous Smithfield foods for both everyday menus and party use. Contains 20-ounce can of James River Smithfield chicken Brunswick stew, one 10-ounce can each of James River Smithfield pork barbecue and turkey barbecue with Smithfield ham, a 10½-ounce can of James River beef barbecue, two 7½-ounce cans of James River Smithfield beef stew, a 7-ounce jar of Amber Brand Deviled Smithfield ham, a 14-ounce jar of James River Smithfield barbecue and meat sauce, two 7½-ounce cans of James River Smithfield chili con carne with beans, a 5-ounce jar of sliced cooked Amber Brand Smithfield ham, and a 4½-ounce jar of James River Smithfield deviled meat spread. $15.95 each prepaid.

SMITHFIELD HAM AND PRODUCTS CO., INC.
Smithfield 8, Va. 23430

GENUINE SMITHFIELD HAMS

Serve a genuine Smithfield ham at your next holiday dinner. Three hundred years of experience stand behind their quality. Cut from peanut-fed porkers, Smithfield's hams are dry-salted, spiced, smoked with hickory, apple, and oak woods, and aged from one to two years. Then they are slowly baked, basted in wine, browned, and garnished in the colonial tradition. Finally, they are vacuum plastic wrapped and gift-boxed, ready to send to you or your friends to carve, serve, and savor. 11–12 pounds per ham—$45.43.

MEAT, FISH AND POULTRY 35

SMITHFIELD'S TREASURE CHEST

For three hundred years, Smithfield has been providing the American public with the finest gourmet and convenience foods. Try this gift assortment and you will want to try all of them. The Treasure Chest contains cocktail slices of cooked Smithfield ham, deviled Smithfield ham, pork barbecue, chicken Brunswick stew, beef barbecue, deviled meat spread, and a bottle of special barbecue and meat sauce. Only $12.90 prepaid.

JAMES RIVER HAM—UNCOOKED

If you want to give a milder flavored ham from Smithfield, select the James River Smithfield hickory-smoked ham. A genuine Smithfield ham, dry-salt-cured, pepper-coated, heavily hickory-smoked, and mellowed to perfection by four to six months of aging. Just right to cut into steaks and broil, cut into slices and fry, or bake by your favorite recipe. Wonderful for making "red eye" gravy to serve with grits or mashed potatoes. These hams come uncooked only. 14–16 pounds per ham—$36.00.

MANHATTAN STEAK COMPANY
148 Greene Street
New York, N.Y. 10012

HAWAIIAN LUAU BROCHETTE

Add new taste to your next outdoor barbecue: with the Hawaiian Luau Brochette sirloin. Made of tender skewered cubes of tasty sirloin with pineapple, this fine meat is specially marinated, Hawaiian style, to give you a tangy taste that will add zest and vigor to your meal. Try it the next time you really want to impress your guests. It's a guaranteed hit!

 Six 8-ounce brochettes $12.25
 Twelve 8-ounce brochettes $22.50

MANHATTAN STEAK COMPANY
148 Greene Street
New York, N.Y. 10012

CHATEAUBRIAND

Here's a cut of beef that's a favorite in France: Chateaubriand. This delicious steak is cut from the very center of the tenderloin—to give you the best quality meat possible. It's succulent, juicy, and tender—perfect for your next gourmet meal. This is the same steak that has been long prized in the great wine-making regions of France—by many connoisseurs of gourmet meats. Serve Chateaubriand at your next holiday feast. Discover its tasty flavor for yourself. Six steaks, 24 ounces each—$59.40.

FIN 'N FEATHER FARM
R.F.D. 2
Dundee, Ill. 60118

SMOKED DUCK

If you go for the exotic in meats, try this delicious smoked duck. Cured by their old-fashioned Fin 'n Feather method (which is exclusively their secret), it's then smoked slowly over hickory fires to bring out its rich, succulent flavor. Here's the perfect entree for those happy holiday meals—the sweet, dark meat is a mouth-watering treat that guests will certainly love. Each duck weighs from 2¼ to 2¾ pounds and comes ready to eat. Price is $8.95 for one duck or $16.95 for two ducks.

ABOUT THIS COMPANY

Fin 'n Feather Farm has been refining the art of curing and smoking for over thirty-five years. You must be completely satisfied, or your money back. Send for their free, full-color catalog.

RATH PACKING COMPANY
Gift Department
P.O. Box 330
Waterloo, Iowa 50704

THE MIDWEST IS GOURMET STEAK COUNTRY

Add to your dining pleasure with gourmet steaks shipped to you directly from the Midwest—beef country. If you enjoy thick, juicy beef, you'll appreciate these fabulous filets. They are cut from aged U. S. Government inspected beef to ensure the most succulent and flavorful steaks ever. These steaks are individually wrapped and quick frozen at their flavor peak, then packed in a reusable Styrofoam container that's ideal for picnics and outings. Eight 6-ounce filet mignon steaks, cut 1¼" thick—$35.00.

For a complete catalog which describes their sirloin strips, T-bones, and other fancy foods, write to Rath's at the above address.

RATH PACKING COMPANY
Gift Department
P.O. Box 330
Waterloo, Iowa 50704

REAL COUNTRY-FLAVORED HAMS

If you yearn for the tasty goodness and moist flavor of real country ham, then a genuine hickory-smoked ham has got to be for you. The Rath hickory-smoked ham, prepared in the best Iowa tradition, has been carefully smoked to retain its own natural flavor. Whether this ham is for you or to be given as a gift, it will be welcomed. The Des Moines is the name given to their 10-pound hickory-smoked canned ham. It costs just $26.50.

For a complete catalog which describes their other country hams and gourmet foods, write to Rath's at the above address.

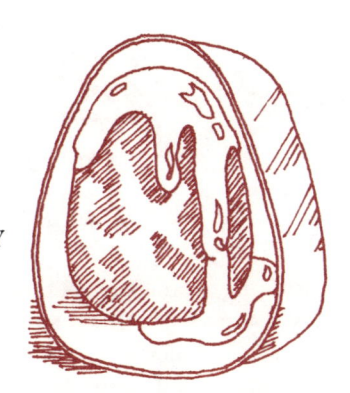

RATH PACKING COMPANY
Gift Department
P.O. Box 330
Waterloo, Iowa 50704

A HONEY OF A HAM

Here's a candied ham that will really tantalize your taste buds. The secret is in the delicious coating of honey, sugar, mustard, and cloves. It's a flavorful sensation that will add to your dining pleasure any time of the year . . . whether it's for a buffet dinner or a family picnic. Available in sizes ranging from 24 ounces to 7 pounds, the most popular is called the Spring Hill. It weighs 5 pounds and costs $15.95.

For a complete catalog which describes their other honey-coated hams and fancy foods, write to Rath's at the above address.

WILLIAM PENN GIFTS
6515 Castor Avenue
Philadelphia, Pa. 19149

SAUSAGE MAKERS' PRIDE

Perfectly seasoned selection of tasty wurstmacher favorites slowly smoked over natural hardwoods for superior flavor. Gift package contains two links each of Landjaegers and Hickory Twigs (excellent sliced for cocktail snacks), 8 ounces tea cervelat, old-fashioned farmer's sausage, all-beef salami, and a giant 2¾-pound all-beef summer sausage. For extra interest, we've added links of sharp Cheddar cheese spread and port wine cheese spread. $28.50 delivered.

See Kitchen Utensils chapter for company description.

EARLY'S HONEY STAND
Rural Route 2, Box 100
Spring Hill, Tenn. 37174

OLD-TIME PORK SAUSAGE

From the first time you catch the aroma of this tangy sausage frying in your pan, you'll know it's not like any you ever ate. Early's makes it from an old family recipe with sage, peppers, herbs, and spices. The only other ingredient is lean, full-of-flavor ground pork. You won't find a speck of cereal or filler in Early's sausage! It's packed in an old-time cloth poke and smoked the same old-time way . . . slow and easy, over green hickory wood. Sizzle up a batch of this sausage with eggs—or put it on a biscuit—or any way you like. (It's even a great treat on pizza.) Once you taste this old-fashioned hill country sausage, you'll know that making really good sausage is not a lost art. A poke weighs at least 3 pounds. Makes a great, out-of-the-ordinary gift. Costs about $1.80 a pound plus shipping. Shipped only November 1 to March 20. Order "Sausage." For each poke, send: Zone A: $6.60; Zone B: $7.10 (postpaid). For each additional poke to the same address, send: Zone A: $5.85; Zone B: $6.35 (postpaid).

MEAT, FISH AND POULTRY

DRY-SUGAR-CURED HICKORY-SMOKED BACON

Any resemblance between Early's bacon and "store-bought" bacon is purely an accident! They still cure it pretty much the same way the pioneers did. Each slab is hand rubbed with honey and dry-sugar-cured in their own mixture that includes special sugars. (That's where the slightly sweeter taste comes from.) Then these lean-streaked slabs get a long, leisurely dose of hickory wood smoke. There are no short cuts—no dipping, needling, or pumping. Your bacon comes to you just as it comes out of their smokehouse—full of hill-country flavor you won't find elsewhere. Shipped year round, since it needs no refrigeration (which makes it great for campers or outdoorsmen). Slabs average about 10 pounds. Price per pound—$1.99. Postpaid prices below include shipping. It's seldom more than this—usually less. If it's less you get a refund for every extra penny. Order "Bacon." For each slab, send: Zone A: $22.50; Zone B: $23.50 (postpaid).

TENNESSEE COUNTRY SMOKED HAMS

It's old-fashioned know-how that gives these hams unmatched flavor. First, they're dry-salt-cured, then they spend months soaking up savory hickory smoke all the way through and flavor-aging in Early's smokehouse. No commercial "tricks of the trade" are used on these hams . . . no quick cures, liquid smoke, needles, or dips! They're prepared the lazy, hill-country way. You can tell when you take your first bite. They run 12 to 16 pounds each. Care and cooking instructions included. Early's guarantees every one to be sound but naturally can't guarantee they'll please every taste. Not everyone likes country ham . . . whether theirs or anybody else's. It is much different from the usual store-bought ham . . . in flavor, firmness, texture, etc. The cooking is a bit different. The hams are cured in salt. Your individual taste will determine whether or not you like aged, smoked country ham. Price per pound—$2.29. Postpaid prices below include shipping. Order "Whole Ham." For each ham, send: Zone A: $32.00 to $39.00; Zone B: $34.00 to $41.00.

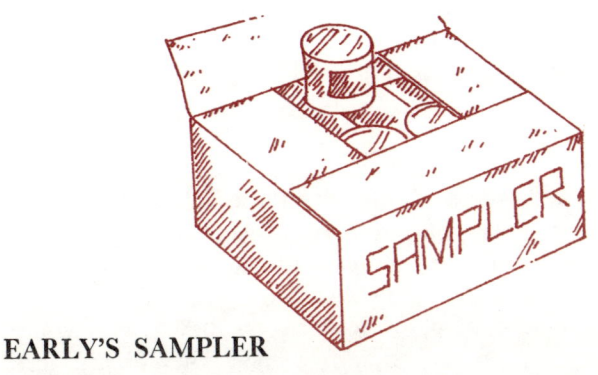

EARLY'S SAMPLER

If you find yourself a little like the kid at the penny candy counter with only one penny, order their Sampler. You get to try some of everything and then decide which you want to come back for more of! Or if you have someone on your list you're not sure of, this is ideal. It includes 2 pounds of their country smoked sausage, four slices of ham, about 2 pounds of hickory-smoked slab bacon, and as a finishing touch, an 8-ounce jar of Early's Pure Bee Honey. Whether you want to treat a friend or do some sampling of your own, you can't go wrong with a Sampler. It's got something for everyone! Order "Sampler." Zone A: $15.95; Zone B: $16.95 (postpaid).

ABOUT THIS COMPANY

It was over fifty years ago that Mrs. Early put some Mason jars of honey on a little "stand" beside a county road running past her farm home, priced them, nailed a tin can to a post for folks to put the money in, went back to her chores, and Early's Honey Stand was started. That fall, the Earlys sold their extra sausage at butchering time along with the honey. The following spring and summer, their surplus hams and bacon went the same way . . . plus all they could buy from the neighbors. And that started them in the mail-order business. Mom and Dad Early are now gone, but the children still carry on the same traditions, ideals, and ways of curing meats that they were taught. They have never changed from the old-time methods. They cure with sugar, salt, honey, and natural spices and herbs, then slowly smoke with green hickory wood. Their meats truly have that old-time flavor and taste. Write for their free catalog, which includes zone information.

HARRINGTON'S
Richmond, Vt. 05477

COB-SMOKED BACON

Here's the Vermont way to start the day. Just pass by the kitchen and you'll smell the difference. Delicately smoked with corncobs and maple wood, every bite is lean, tender, and smoky sweet. It tastes the way that bacon should taste. Your breakfasts will take on new meaning. 2 pounds (sliced)—$6.65. 5 pounds (sliced)—$14.40.

ABOUT THIS COMPANY

Harrington's is intrinsically New England—Yankee to the core, you might say. They believe strongly in the basic virtues that are part of their heritage: honesty, integrity, thrift, simplicity, and a wonderful ability to make do. For over one hundred years, they have provided the New England housewife with foods that are a basic part of the New England diet, and in recent years they have extended nationwide through marketing these products by direct mail. They process their own meats and fowl and are federally licensed by the USDA. The method of processing these products is somewhat unique in that they smoke the pork products, turkeys, pheasants, and ducks with corncobs and maple wood, which imparts a very unusual delicate flavor.

MEAT, FISH AND POULTRY 41

BRUSS COMPANY
3548 North Kostner Avenue
Chicago, Ill. 60641

EYE-OF-THE-RIB ROAST

Ready for a really magnificent meal? Try a juicy, succulent eye-of-the-rib beef roast. This extremely tender meat is carefully selected and cut to give you a superb entree that will be the highlight of that important dinner. It's taken from high-quality U.S. Prime beef—with backbone, short ribs, and extra fat removed, so you receive only the best selection of meat. Serve eye-of-the-rib beef roast at your next supper party or holiday dinner. It's a guaranteed success! Each roast weighs 11–13 pounds and costs $56.50.

SCHALLER & WEBER, INC.
22–36 46th Street
Long Island City, N.Y. 11105

WESTPHALIAN SMOKED HAM

Do you enjoy a gourmet meat buffet? Then why not sample an Old World Westphalian smoked ham? This mild (ready-to-eat) cured ham is a delicious addition to any dinner or midnight snack. It's cooked to perfection—with just the right touch of spice to make your mouth water. When you taste it, you'll understand why the manufacturer has consistently received the International Exhibition's Gold Medal of Honor.

Choose a traditional Westphalian smoked ham, in 2-pound, 3-pound, 4-pound, and 5-pound sizes at $4.19 per pound. Or taste the Nuss-Schinken, a smaller-sized Westphalian, in 3- or 4-pound sizes at $4.19 per pound. The Bauernschinken is a flat-shaped Westphalian Ham which runs from 5 to 8 pounds and also costs $4.19 per pound. The Schinkenspeck is also delicious. It runs from 2 pounds up and costs $2.98 per pound. It should be sliced thin and served as an appetizer.

You can also take your pick from Schaller & Weber's fine assortment of liverwursts, salamis, other smoked hams, bolognas, sausage, and other fine meats. For a free catalog and price list, write the above address.

V. W. JOYNER & CO.
Main Street
Smithfield, Va. 23430

RED EYE HAM

If country ham is a favorite around your house, you'll want to try red eye Virginia country ham. Made with an exclusive southern formula, this plump, juicy ham is pepper-coated and carefully aged in the best Virginia tradition. And the red eye ham comes already baked and glazed, so you can slice and serve it anytime. It's the perfect food for that holiday dinner, summer picnic, or evening buffet. Joyner's red eye Virginia country ham costs just $2.90 per pound (postage included) and is available in sizes from 9 to 12 pounds.

EPICURES' CLUB
Elizabeth, N.J. 07207

SMOKED PHEASANT

Looking for a super entree to serve your guests at holiday time? Here it is: the smoked pheasant. Carefully cured and smoked over hickory embers, this delightful bird is the perfect main dish for any gourmet meal. And it's specially canned, in delicious pheasant broth, to preserve that high-quality natural taste. The smoked pheasant can be eaten immediately or saved for future occasions. Serve it at your next buffet supper, Thanksgiving dinner, or anytime you want to make a distinctly flavorful impression. A 3-pound pheasant costs just $10.95. Or order a special gift box containing pheasant, mixed nuts, Portuguese sardines, orange slices, plum pudding, midget pickles, Spanish olives, and rainbow trout pâté—all for only $16.50.

ABOUT THIS COMPANY

During the twenty-seven years they've been in business, many millions of packages have been shipped from their plant. Over 800,000 people and companies have entrusted the handling of their gift orders. Every one of those packages and every one of those orders were backed by a policy of unconditional satisfaction guaranteed. But seeing is believing—and they'd like you to see these things for yourself. This is a cordial invitation to visit them whenever you're in the neighborhood. (Take Route 22 to Union, New Jersey, turn off onto Route 82 in the direction of Elizabeth; go 1.8 miles to Lehigh Ave. Turn right, go 500 feet, and you'll be at their door.) The welcome mat is always out.

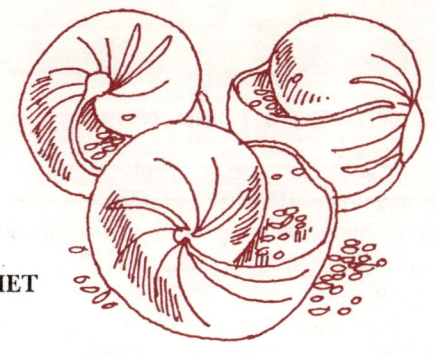

LE JARDIN DU GOURMET
Raymond Saufroy, Imports
West Danville, Vt. 05873

ESCARGOTS

One way to really dress up a meal is to serve escargots—those delicious little snails. Whether you serve them as an appetizer or main course, make it a special event by having all the equipment you need. Snails can be ordered by the tin.

Tin of 12	$2.10
Tin of 24	$4.10
Tin of 48	$8.25

Shells for snails are $1.10 per dozen. These shells can be washed and reused. Snails should be served on special stainless-steel plates. 6-holed plate—$5.85. 12-holed plate—$7.95. Stainless-steel snail pincers are $1.75 each and stainless-steel snail forks are $.85 each.

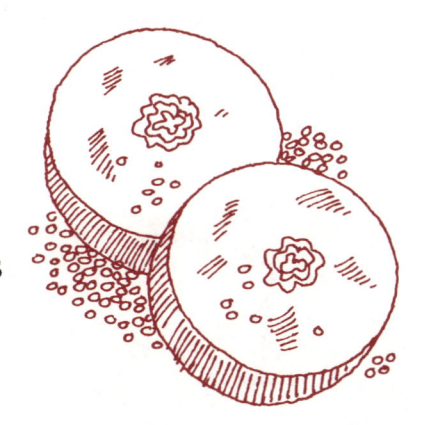

EMBASSY SEAFOODS
P.O. Box 268
3 Cottage Park Road
Winthrop, Mass. 02152

CODFISH CAKES

The heart of sturdy Yankee breakfasts . . . and an absolute must for Saturday night supper! Embassy has combined salt cod fluff with other choice ingredients to make an authentic New England codfish cake mix. Each 10½-ounce can makes from four to six generous cakes. Use as is, add it to your corn-fritter batter, create your own special dishes. Its uses are almost unlimited! Three 10½-ounce cans—$3.50.

BRUNO'S PEN & PENCIL
205 East 45th Street
New York, N.Y. 10017

SIRLOIN STEAK AND FILET MIGNON

On Manhattan's East Side, there is a famous restaurant known for its incomparable Prime beef. Residents and visitors go there by the droves to taste of this delicacy. Now you and your guests can enjoy the same high quality that has made Bruno's world-famous. Each sirloin steak weighs 12 ounces and is 1½" thick. Each filet mignon is 8 ounces and is 1¾" thick. Chosen from the same beef which they serve their patrons, this meat is unsurpassed. Cost of the sirloin steak (6 to a pack) is $60.00. Cost of the filet (6 to a pack) is $50.00. Add $5.00 for shipping costs for one or two items; $10.00 for shipping on three or more. Add $5.00 for air shipping.

EMBASSY SEAFOODS
P.O. Box 268
3 Cottage Park Road
Winthrop, Mass. 02152

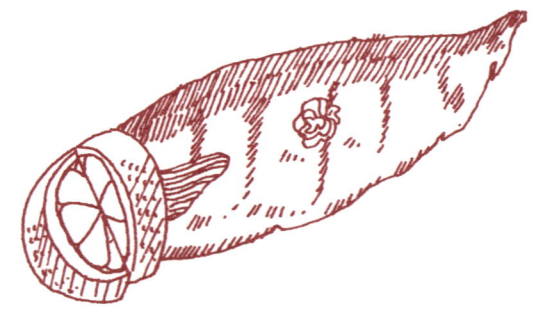

SALT MACKEREL

The aromatic flavor of this historic delicacy brings a refreshing ocean breeze to your kitchen. Each one of the plump mackerel weighs nearly a pound and provides a gourmet treat for three people. Broil, bake, or boil to suite your taste. Selected choice fillets are packed in a reusable wooden bucket with brine and pure white table salt. Instructions for freshening are on the lid. 5-pound pail—$12.00. 10-pound pail—$24.00.

EMBASSY SEAFOODS
P.O. Box 268
3 Cottage Park Road
Winthrop, Mass. 02152

SHAD ROE

If you're a connoisseur of fine seafood, you'll be happy to know that you can buy the finest obtainable shad roes by mail. These high-quality roes are carefully chosen for excellent quality and condition, then partially precooked with a seasoning of salt to preserve their ocean-fresh sea flavor. Each can contains a full 7¾ ounces. Broil with bacon to a golden brown and garnish with lemon and parsley, to serve one of the best seafood dishes imaginable. Three 7¾-ounce cans—$17.50.

MEAT, FISH AND POULTRY

EMBASSY SEAFOODS
P.O. Box 268
3 Cottage Park Road
Winthrop, Mass. 02152

SALT COD

Cured in the traditional way, handed down among seafarers for three hundred years, this "just caught" goodness is passed to you. Each tasty fillet has the fresh, natural flavor of cod preserved from the North Atlantic fishing fleet. If you've been looking for a new main course to serve at your family supper, holiday buffet, or late-evening dinner, look no further. With salt cod fillets, you're sure to have a winning meal. Each wooden pail contains 5 pounds net weight of choice fillets of salt cod. 5-pound pail—$15.00.

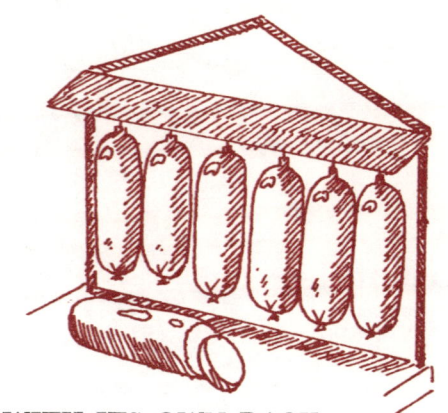

FIGI'S, INC.
Marshfield, Wis. 54449

A VARIETY OF SAUSAGES WITH ITS OWN RACK

A delightful treat for the family! A unique gift for sausage-loving friends! A combination hanging rack and hardwood cutting-serving board with six interesting smoked sausages: all-beef summer sausage, beerwurst, braunschweiger, extra-lean Black Jack summer sausage, metwurst, and pizza sausage. Put the rack in the kitchen for quick snacks, or bring it into the living room to serve with cocktails. Great for informal parties and around the pool. $15.95 delivered.

See Cheese chapter for company description.

FIGI'S, INC.
Marshfield, Wis. 54449

SMOKED TURKEY BREAST

Full-flavored smoked turkey breast, hardwood-smoked to a luscious brown for "right out of the oven" country flavor and tenderness. Fully cooked and ready to eat. Net weight 4–5 pounds. Available September through April. $15.95.

See Cheese chapter for company description.

FIGI'S, INC.
Marshfield, Wis. 54449

REAL COUNTRY BACON

Succulent, lean slab bacon—slowly sweet-cured to perfection in the old-fashioned way, ready to slice to suit individual preferences in thickness. A delicious day-brightener at breakfast; a tasty treat when incorporated in other meals. Slab bacon is available in three different sizes.

2 pounds	$ 7.50
4 pounds	$12.50
6 pounds	$15.95

See Cheese chapter for company description.

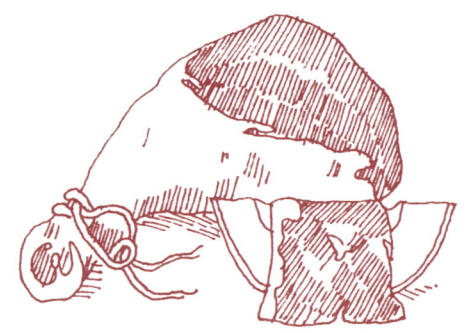

MANGANARO FOODS
488 Ninth Avenue
New York, N.Y. 10018

IMPORTED PROSCIUTTO HAMS, ETC.

Prosciutto A truly fine and exquisite cured ham. Its delicate flavor makes it a treat on all occasions. Serve paper thin. Delicious to eat with ripe figs, melon, or rolled around bread sticks. Boned. In bulk (5 pounds average)—$7.50 per pound.

Coppa di Piacenza Style A mild, soft, and aromatic cured ham, carefully trimmed and gently placed in its own natural casing. Domestic. In bulk (4 pounds average)—$4.80 per pound.

Mortadella A delicately prepared "cold cut" from Bologna, and shaped like bologna, spiced with rare herbs and a touch of garlic. Serve on antipasto or cold-cut tray. Makes tasty sandwiches. $3.50 per pound.

Prosciuttino A popular luncheon meat of carefully cured ham, without fat, seasoned by generous peppering. Domestic only. Boned. In bulk (4 pounds average)—$4.60 per pound.

Pancetta An Italian-style cooking bacon, adds fine flavor when cooking fresh peas, minestra chowders, etc. In bulk (4 pounds average)—$4.60 per pound.

See Special Foods chapter for company description.

MEAT, FISH AND POULTRY

MANGANARO FOODS
488 Ninth Avenue
New York, N.Y. 10018

ITALIAN SALAMIS

Citterio Brand   A finely ground pure pork, spiced lightly with garlic and black pepper. Assorted sizes from 1 pound up. Imported. $6.50 per pound.

Sopressata   A salami style with more marbling but no garlic. Slow-cured highly aromatic pork in natural casing. Chewy texture, should be cut in ¼″-thick slices to be eaten out-of-hand with bread. $4.80 per pound.

Casalinga Style   A flavorful domestic version. Pure pork ground in large chunks, seasoned with garlic and black pepper, cured in an age-old tradition, and smoked by charred wood. Made in 1-pound size. $4.80 per pound.

See Special Foods chapter for company description.

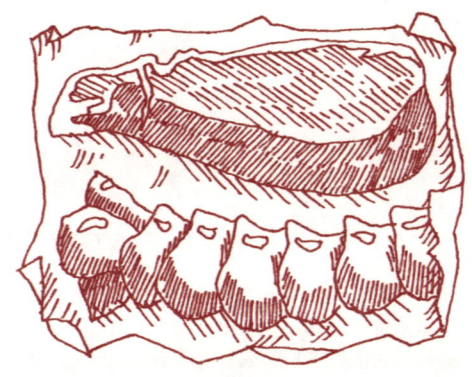

PFAELZER BROTHERS
4501 West District Boulevard
Chicago, Ill. 60632

SURF & TURF

The kings of meat and seafood are united to form a gift package even your mother-in-law will be enamored with. Prime filet mignon, cut from the center of the tenderloin, and South African lobster tails give your recipients a meal heretofore offered in only the world's finest restaurants—Surf & Turf. This is the gift selection for those you really want to impress. 5 filets mignons, each 7 ounces, and 5 lobster tails, each 10 ounces—$50.00.

ABOUT THIS COMPANY
In 1923 the Pfaelzer Brothers went into the meat business with a single, clear-cut purpose—to provide the finest meat, poultry, and seafood that money could buy. They deserve their hard-earned reputation. Send for their free, full-color catalog.

PACKING SHED
P.O. Box 11
Weyers Cave, Va. 24486

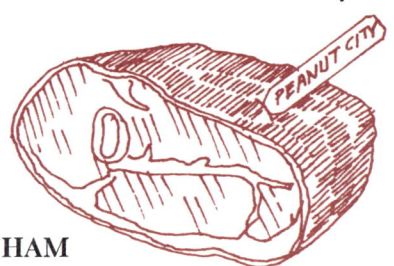

PEANUT CITY COUNTRY HAM

Now—for those who are tired of tough, dry, too-salty ham—here's a refreshing idea: "Peanut City" country ham. One of Virginia's traditional foods from the historic tidewater region, this 10–12-pound short shank ham is 100 per cent dry-cured in salt, smoked very slowly, and carefully aged for two to three months. This is the same slow cure used by the Virginia colonists. Each ham is USDA-inspected, and is uncooked. Mild in flavor, this country ham does not require soaking or other elaborate preparation. Each 10–12-pound ham costs just $24.45, and comes with complete cooking instructions.

AMANA SOCIETY MEAT SHOP
Amana, Iowa 52203

AN ASSORTMENT OF SMOKED FAVORITES

Here's a package of fine, hand-selected meats you'll long remember. A light-smoked and double-smoked summer sausage, each containing its own special blend of fine ingredients and rare imported spices, then carefully smoked to bring out its full flavor. The bacon takes nearly a month to prepare. Chosen only from Prime Iowa corn-fed hogs, it's processed to perfection in Old World tradition. The ham, pretty as a picture, is cured in a special brine, then hung and smoked. And the smoked Cheddar is a rare treat in itself. 5–6 pounds—$16.95.

ABOUT THIS COMPANY

You are invited to visit the Amana Meat Shop in Amana, Iowa! Here is an interesting old-time butcher shop with a towering smokehouse built in 1858. You can browse and sample their fine products. Next time you're in the Amanas treat yourself to an adventure in good eating. They'd love to have you! The story of Amana began about 250 years ago in Germany when a group seeking religious freedom migrated to America. In this group were master butchers bringing their recipes and "trade secrets" with them. As a result of the scrupulous maintenance of quality, these distinctive smoked meat specialties are now gracing tables everywhere. The Amana conviction is that time itself is a vital ingredient. Curing is still done in the slow, old-fashioned process. The excellence of Amana meat products is a time-honored tradition.

AMANA SOCIETY MEAT SHOP
Amana, Iowa 52203

WORLD-FAMOUS SMOKED HAM

These hams are cured with a special piquant brine. After the curing process, the hams are hung in a century-old smoke tower where they are bathed in blue-gray waves of woodsy smoke until each ham reaches a fine point of truly superb flavor that's never been equaled. After completion of this skillful process, the hams are fully cooked, ready-to-eat, and prepared for shipment. They are individually wrapped and hand-packed in sturdy, attractive gift cartons. Amana hams are truly a gourmet's dream come true. 14–15 pounds—$34.95.

See company description in this chapter.

AMANA SOCIETY MEAT SHOP
Amana, Iowa 52203

BIG STICK SUMMER SAUSAGE

Are you tired of picnics with nothing but hamburgers and hot dogs? Then try this new favorite: Big Stick summer sausage. A special blend of spices is added to bring out the delicious flavor of the specially selected meat. Then they spend several days lazily soaking up flavor in thick clouds of smoke from smoldering embers. This succulent sausage comes fully cooked. Superbly seasoned and smoked, it's almost two feet of the finest ingredients available—ready to eat anytime. Serve it for a light Saturday brunch, on a picnic lunch, for an outdoor supper after a long day of hiking, or even as a midnight snack with cheese. However, whenever you eat it, you'll love its hearty, zesty flavor. Each sausage weighs 3 pounds and costs $9.95.

See company description in this chapter.

GOKEYS
21 West Fifth Street
St. Paul, Minn. 55102

BRACE OF BIRDS

Imagine what a succulent meal you could have with two mallard ducks or two pheasants. These excellent birds are game farm raised, and are NPIP certified and tested. Each bird weighs between 2 and 2½ pounds. They ship your "Brace of Birds" packaged in dry ice via UPS to Minnesota and bordering states and via air mail to all others, at an additional charge of $4.00 for the first brace and $2.00 for each additional brace.

    Mallard "Brace of Birds" $32.50
    Pheasant "Brace of Birds" $32.50

PFAELZER BROTHERS
4501 West District Boulevard
Chicago, Ill. 60632

LOBSTER TAILS

Here, from the waters off the coast of South Africa, are lobster tails that are world-famous for their snow-white, juicy meat. An elegant gift because of their mouth-watering flavor, or a tantalizing treat for yourself. Whether served broiled with butter, in spectacular salads, or at barbecues on skewers, these tails offer a meal that's always exciting and interesting.

    8 lobster tails, each 10 ounces $49.50
    6 lobster tails, each 12 ounces $47.00

    See company description in this chapter.

FRUITS AND VEGETABLES

Because fruits and vegetables are so abundant across the country, people tend to take them for granted. Too often, Americans just don't come in contact with what must be considered the crème de la crème of this produce. For superlative fruits and vegetables can only be considered as one of the best eating experiences in the world.

For example, there is an American pear which was originally developed in southern France in the mid-nineteenth century. Introduced into America in the late 1800s, today it flourishes in the warm fertile valleys of California, to be enjoyed by connoisseurs throughout the land.

Oranges, miracle grapefruits, and Lisbon lemons are not only packed with vitamin C, but they provide delicious eating, too. Hand-picked, they are an unforgettable taste experience. And can you imagine what it's like to cut open a luscious, naturally sweet grapefruit which weighs nearly a pound? Or a crunchy apple which literally drips with mountain-grown goodness?

Fresh, crisp vegetables don't have to be a thing of your imagination. Every part of the nation is ready and willing to ship you their specialties. From artichokes to zucchini and other delicacies, you can have them all. Choose from any number of quality mail-order houses which specialize in shipping you the highest quality fruits and vegetables.

HALE INDIAN RIVER GROVES
Wabasso, Fla. 32970

TROPICAL FRUIT CLUB

Give yourself or your friends a year-round gift—with a rich assortment of fruits and jams from the Tropical Fruit Club. These luscious treats are carefully selected and freshly packaged—to preserve their zesty taste. Each month brings you a different delicacy: red tangelos, navel oranges, mangos, avocados, orange blossom honey, seedless grapefruit, and many other varieties. Choose from a three-, six-, nine-, or twelve-month club—any way you take it, the Tropical Fruit Club will fill your house with flavor.

Three-month	$25.00
Six-month	$48.00
Nine-month	$69.75
Twelve-month	$89.95

AGWAY, INC.
Box 5030
Syracuse, N.Y. 13250

APPLE PACK

Do you have a hard time finding really top-quality apples? Then you should send for the Apple Pack. Carefully selected and packaged by hand, these juicy Red Delicious apples are among the finest money can buy. And because they're shipped to you immediately, they look, smell, and taste like they just came off the tree. Attractively packaged in a 16-apple carton, the Apple Pack is also an ideal gift. Cost is just $6.95, including shipping.

MISSION PAK
Santa Clara, Calif. 95050

GLITTERING GLACÉED FRUIT

During the middle 1920s, George Page, founder of Mission Pak, traveled widely in Europe and chanced on something in France that was to become Mission Pak's greatest success. It was the glacé process. In Nice he watched French cooks glacé fresh fruit in stone crocks. They would not reveal their formula for this marvelous process, so George Page returned to the States and began to experiment, finally turning out a glacé process far superior to the original French method. Instead of just covering the fruit with a thick sugar syrup coating, he developed a technique which replaces the moisture inside with orange blossom honey. When all the water content is removed from these rare delicacies, each piece is simmered till just a mist of sweetness surrounds it—blending with the rich natural flavor of freshly picked fruit. Individually packed, piece by piece, it includes plump Deglet Noor dates, luscious moist pineapple, bright red cherries, sliced pears, delectable prunes, tender, tasty apple slices, firm, tempting whole apricots, delicious Calimyrna figs, tangy orange slices. You may remember your first taste of glacéed fruit as a youngster. It's been a traditional Christmastime treat in many families for years. The taste will never be forgotten. Net weight 4 pounds. $14.95 delivered.

WILLIAM PENN GIFTS
6515 Castor Avenue
Philadelphia, Pa. 19149

ASSORTED FRUIT AND CHEESE BOX COMBO

For a delicious combination of fruit and cheese, try William Penn's attractively packaged Fruit and Cheese Box. It's filled with fourteen of the most luscious apples, pears, and oranges available—as well as four different kinds of gourmet cheeses: brick, Edam, aged Cheddar, and mild Cheddar. The Fruit and Cheese Box is a marvelous mixture that will delight both you and your guests. Keep it around whenever company drops by (they'll love you for it), or give it as a holiday gift (it's the perfect present). The Fruit and Cheese Box weighs 10 pounds and costs $13.95 delivered.

See Kitchen Utensils chapter for company description.

R. H. CHAMBERLIN
Gift Fruit Shipper
P.O. Box 87
Sharpes, Fla. 32957

FLORIDA CITRUS AND SEAFOOD

Can't get down to Florida this year? Want to sample Florida citrus or seafood before deciding to vacation there? Need some unique gift ideas? R. H. Chamberlin offers you the finest of Florida's foods shipped direct to your friends or your home. Juicy luscious Florida oranges, grapefruits, tangerines, limes, mangos, coconuts, and avocados. Picked at the pinnacle of perfection. No artificial coloring or flavor.

And what about Florida's world-famous seafood? Stone crab claws, blue crab claws, crabmeat dressing, lobster, rock shrimp, and red snapper. All ready to cook and serve. Also available in Shore Dinner Combinations for six people, and a Surf & Turf package for four, including Prime filet mignon. Contact R. H. Chamberlin at the above address to learn how you can order these Florida treats. Cost of the Surf & Turf is $41.95; for the All-Fruit Gift Box (1 bushel), it's only $17.95.

MISSION PAK
Santa Clara, Calif. 95050

SUN-DRIED FRUIT

This new and different pack of delicious sun-dried fruit is the perfect gift at Christmastime. The pretty bell-shaped tray is packed full of irresistible dried fruit with all their natural goodness sealed in . . . figs, dates, prunes, almonds and walnuts, apricots, pears, cherries, pineapple, and the popular citrus peel. This is such an appropriate gift for those you wish to remember in a special way. A mouth-watering assortment of dried fruit that will provide never-to-be-forgotten eating pleasure wherever it goes. Net weight 2 pounds. Available year round. $8.95 delivered.

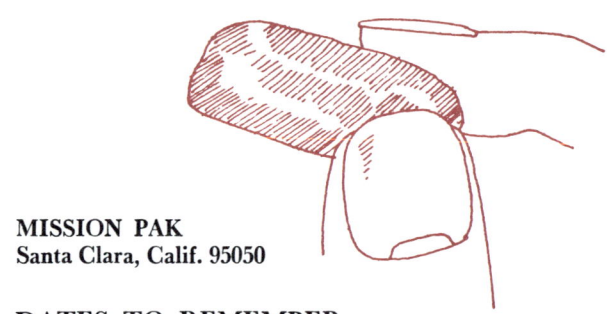

MISSION PAK
Santa Clara, Calif. 95050

DATES TO REMEMBER

Delicious Deglet Noor dates are naturally ripened on giant date palm trees in the desert valleys of California. Warm days, bright sunshine, and cool crisp nights slowly turn each date sugar-ripe. Deglet Noor dates are Mother Nature's own confections: sweet in their natural sugar, an energy food as old as time, popular and healthful. These fantastic dates are picked and packed at the peak of plump ripeness, so fresh and tender they literally melt in your mouth. Some are decorated in silver foil, others are stuffed with meaty giant walnuts for the ultimate in eye and taste appeal. Truly an irresistible treat for any occasion. Available the year round.

2 pounds	$ 6.95 delivered
3 pounds	$ 9.95 delivered
4 pounds	$11.95 delivered

FRUITS AND VEGETABLES

MISSION PAK
Santa Clara, Calif. 95050

BASKET O' SUNSHINE

Here's a sweet, juicy way to brighten up any holiday: bright navel oranges from California. These huge, succulent fruits are the perfect addition to your table centerpiece or Christmas stocking. They're grown in the fertile soil and climate of warm California, then picked and packaged immediately—so you get the benefit of their full-flavor taste. And the oranges are attractively packaged in a woven basket—with orange marmalade, dates, and walnuts tucked inside. Fill your next holiday with a Basket o' Sunshine. It's a gift that will be used long after its contents are gone. Each basket weighs 13 pounds and costs $13.95.

WILLIAMS-SONOMA
Mail Order Department
532 Sutter Street
San Francisco, Calif. 94102

CHINESE BEAN SPROUTS

Even the most inexperienced gardener cannot fail to grow this delicate Oriental vegetable right in the kitchen. Suttons Chinese bean sprouts (mung beans) germinate in six to nine days and need no soil; just a damp surface, such as cotton wool, a tray, and a dark corner. These bean sprouts make a delicious addition to any Oriental food, and are equally good in almost any kind of sandwich. The taste is reminiscent of tiny new raw peas; very delicate and sweet. Suttons are seedsmen to the British royal family, so if they are satisfied, we think you will be too! Two packs, 8 ounces each—$4.50.

See Kitchen Utensils chapter for company description.

LE JARDIN DU GOURMET
Raymond Saufroy, Imports
West Danville, Vt. 05873

JERUSALEM ARTICHOKES

People are often puzzled as to how you use these little vegetables. And even people who are familiar with them aren't aware of their many uses. You can prepare them any way you prepare potatoes. Even eat them raw, cut in small cubes, in salads. Their taste is akin to that of the artichoke. The oldest way to eat them is baked in the oven. When boiled in salted water, they may be eaten with a vinaigrette sauce—or make fritters. When a recipe asks for artichoke bottoms, you can use the Jerusalem artichoke instead. They are very good in potato purée or mashed potatoes; use 30 per cent artichoke and 70 per cent potato. When your order arrives and they look a little wrinkled, soak them in water twenty-four hours before planting them. Plant them a foot apart in rows one and a half feet apart. They make a stalk over six feet tall and look like sunflowers. They are of the same family. 12 ounces—$2.00 postpaid.

MR. ARTICHOKE
11000 Blackie Road
Castroville, Calif. 95012

ARTICHOKES

Are you tired of paying premium prices for old, tough, woody artichokes? Now you can have the choicest of the crop packaged field-fresh. These artichoke buds are picked from the fields no sooner than the day preceding shipment. You can order a carton of 12 large or 24 small artichokes. Both sizes weigh about 11 pounds each, and are the same price. All artichokes carry an unconditional guarantee of excellence. Enjoy these fresh gourmet treats in salads, pickled, or boiled and served with mayonnaise, drawn butter, or hollandaise sauce. A treat for yourself. A unique gift for gourmet friends. 12 large or 24 small—$9.00 west of the Mississippi; $12.50 east of the Mississippi.

ABOUT THIS COMPANY
Mr. Artichoke is located in the artichoke capital of America, near Monterey Bay, California. Their artichokes are available in two sizes, as described above, and they will ship anywhere in the world, with their guarantee intact.

FRUITS AND VEGETABLES
59

GRACE A. RUSH
3715 Madison Road
Cincinnati, Ohio 45209

MEDJOOL DATES

You're in for a rare epicurean treat if you've never before had Medjool dates. Noted for their flavor and perfection as well as for their astonishing size, Medjools predate the Christian era, and until this century, came only from a stand of palms at a single oasis in Morocco. In 1927 the Moroccan Government was persuaded to send twelve shoots to the U. S. Department of Agriculture in California for test planting. The precious shoots were divided among eleven experienced date growers, one of whom was successful at developing the palms. Today the only remaining stand of Medjool dates in the world is in that small area. (A pestilence several years ago destroyed the entire Moroccan stand.) They are among the world's most pampered fruits. Ninety per cent of the dates on each palm are thinned out, and the remaining 10 per cent are allowed to reach maturity. During the ripening period, the bunches are shielded by paper bags to prevent damage from insects and rain. Picked only when judged by experts to have reached the very peak of development, the dates are shipped immediately to Grace Rush by carefully controlled transit. Their production is limited, and owing to their prized status among connoisseurs of fine food, Medjool dates are available only at one season of the year. 2-pound tin—$9.00.

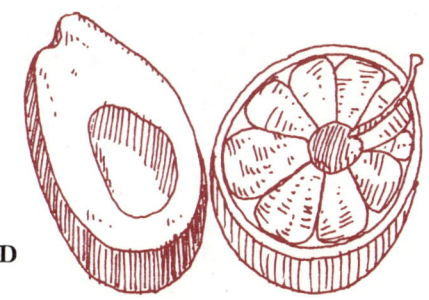

HARRY AND DAVID
Medford, Oreg. 97501

GRAPEFRUIT AND AVOCADOS

It is with great pleasure that Harry and David are again able to offer you the unique holiday gift of grapefruit and avocados. This has been a cherished gift for the holiday hostess, relatives, friends, and for the person with a yen for the best. Succulent grapefruits enhanced by the uncommon flavor of the versatile avocado. All are hand-picked for their quality. Remember, when you give an avocado, you also give a plant. Serving suggestions and information on how to plant the seeds are included. Six grapefruit and six avocados—only $13.95 postpaid.

See Confections chapter for company description.

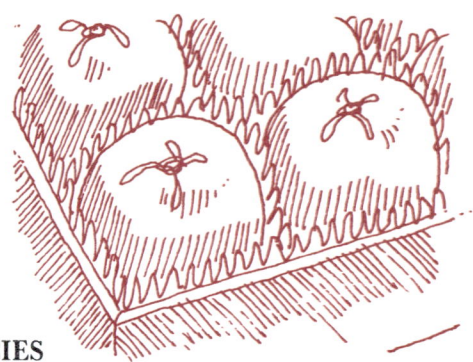

HARRY AND DAVID
Medford, Oreg. 97501

ORCHARD DELICACIES

If you love gourmet fruits, here's a selection that's bound to win your favor. It's a magnificent harvest of plump and juicy Royal Riviera pears, flavorful Royal oranges, and snappy Crisp Mountain apples. These hand-selected beauties are available just when fruit is at its scarcest—during the winter months. There are over 13 pounds of naturally sweet and juicy eating. And because they are boxed individually, you can give them as gifts (if you can bear to part with them). $15.95 delivered.

See Confections chapter for company description.

FRANK LEWIS
P.O. Box 517
Alamo, Tex. 78516

ROYAL RUBY RED GRAPEFRUIT

If you go nuts over great grapefruit, here's an exciting treat for you: world-famous "Miracle" Royal Ruby Red grapefruit. Here's the story. Back in 1929, one of the men who were picking fruit came up to the house holding six of the strangest grapefruit anyone had ever seen! A single branch of an ordinary grapefruit tree had produced these six unusual fruit. These were big grapefruit, unusually big. And they had a faint red blush on their skin. The amazing thing was that when the grapefruit was sliced open, the fruit was a *brilliant ruby red* in color. Today, these plump delectable grapefruit are skillfully grown and fully ripened, then shipped to you orchard fresh to preserve their exquisitely delicious taste. Weighing a pound or more each, these luscious fresh fruits are so sweet they'll never need sugar. Serve them the next time you have company—your guests will rave over their rich, juicy flavor. And the Royal Ruby Red grapefruit is available in a variety of gift packages. The Merry Christmas Medley, for instance, costs $11.95, and contains seven grapefruit, Medjool dates, hard candies, and pecans.

FRUITS AND VEGETABLES

ABOUT THIS COMPANY
Frank Lewis started this company with a grafted branch from a "miracle" grapefruit tree. Today, they specialize in only the finest fruits, all with a complete satisfaction guarantee.

FRANK LEWIS
P.O. Box 517
Alamo, Tex. 78516

FAMILY PACK GRAPEFRUITS

Here is the ideal fruit pack just for your own family and close relatives . . . for economical eating all through the growing season. These are the same first-quality, tree-ripened, full-flavored Ruby Reds as the world-famous "miracle" gift-size fruit, but they are definitely smaller than their big brothers. This Family Pack is more for keeping than for giving. These Ruby Reds have been pampered and cared for just like all their grapefruit . . . only they just didn't grow quite as big. They're every bit as colorful, juicy, and flavorful as the best of the crop. Actually, many people find they are a perfect size for the family at breakfast. Your family will be bright-eyed and happy each morning as they start the day with this vitamin-rich, gleaming, red-meated fruit. So modestly priced . . . and such good eating. Contains 18 medium-sized Ruby Reds, each weighing almost a pound apiece. Available November 15 through April 15. $11.95 delivered.

See company description in this chapter.

VERMONT COUNTRY STORE
Weston, Vt. 05161

OLD-TIME BAKED BEANS

Years ago, the Vermont Country Store introduced their own brand of oven-baked beans, and the New York *Times* proclaimed them "first rate and of excellent flavor." This is because, whilemost canned beans are steamed (not baked), theirs are baked fourteen hours in a brick oven to make them toothsome, mellow, and tasty. In addition, their baker uses real salt pork and typical New England style sauce.

Yellow Eye Beans with Pork are large beans, with a yellow spot or eye. Good meal for three persons. Two 28-ounce tins—$2.25.

Soldier Beans with Pork are also big beans, with brown markings, raised down east in Maine. Soldiers fought on them! Two 28-ounce tins—$2.25.

Jacob's Cattle Beans with Pork are large delicious beans named from Jacob's cattle that the Bible says were "spotted and speckled." So are the beans. They are fine-grain, mellow, and so tender that they almost melt in the mouth. Two 28-ounce tins—$2.25.

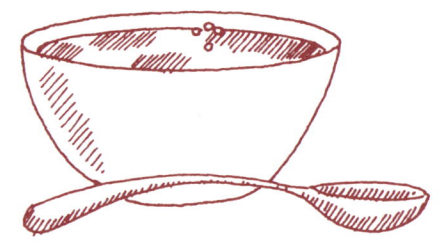

CORN CHOWDER

If anything is evocative of rural New England, it is this tasty, nourishing chowder. Two 15½-ounce tins—$1.65.

PREPARED CLAM CHOWDER

Made with lots of Maine clams and potatoes, theirs is the best. Add milk and serve. New England style, of course! Two 15-ounce tins—$1.89.

DOWN EAST FISH CHOWDER

This chowder is prepared the old-fashioned way: generous hunks of cooked white fish, potatoes, and a touch of onions. Add equal volume of milk, some butter, heat, and serve hot with crackers. Two 15-ounce tins—$1.89.

CHEESE

Originally, making cheese was a simple way of using up excess milk. Today, this product has evolved into a multitude of varieties, many of which are associated with small local regions in Europe and other parts of the world.

Believe it or not, most cheeses are fermented, which means that they have been made from raw curds. A few cheeses are produced by "spontaneous coagulation." They are known as fresh cheese.

Learning how to serve the proper cheeses, and in the correct manner, can take months of instruction and years of practice. In France, traditionally, the cheese is served before dessert, or instead of dessert. Fresh butter may be served with cheese, but the real gourmet would decline to use it.

Cheese has many different purposes. Aside from using it before the final course, it is also used as a dessert, in main courses, and in nearly any type of cooking.

Cheese has been with us for centuries, and it comes in countless varieties. Choose from hard cheeses, firm cheeses, semifirm, semisoft, and soft cheeses. And the tastes are as infinite as the types of cheeses. Choose from sharp Cheddars, mild Cheddars, all the way to the delicate creams which are personified by the soft cheeses.

The various climates which exist around the world dramatically affect the way a cheese tastes. This is because the vegetation which an animal eats affects the flavoring of its milk. Add to this various herbs and spices, and you can begin to see how you can change the cheese flavoring process.

Some firms specialize in a few select hard local cheeses. Others make up unique blends. Still others import all types from around the world and offer the widest selection of fine cheese imaginable (except for fresh country cheeses like cottage cheese or ricotta that need constant refrigeration).

When you order cheese through the mail, you enjoy the convenience of "armchair" shopping, have an enormous variety from which to choose, and the unending pleasure of discovery.

FIGI'S, INC.
Marshfield, Wis. 54449

KAVE KURE CHEESE SPREAD

An exclusive blend of natural aged cheese gives the distinctive taste appeal to this sharp Cheddar Kave Kure. It's a perfect example of the real cheese maker's ingenuity. This tangy, smooth-spreading cheese is well deserving of its ever-growing popularity. Handsomely packed in seal-tight earthenware crocks, here is fully aged cheese to pamper your own discerning taste. 20 ounces—$5.75 delivered. 2½ pounds—$8.50 delivered.

A GIANT BOX OF CHEESE

Here's a legendary box of cheese samplers that would thrill any connoisseur. It contains 5¼-ounce squares each of Cheddar, caraway, and salami; a 4½-ounce square of Swiss; two 4-ounce angles of brick; 2 ounces each of Cheddar, colby, caraway, pepper, smoked Edam, brick, Edam, fiesta, bluette, sharpy, and napoli; four 1-ounce cheese links; and four ⅔-ounce wedges. $9.95 delivered.

See company description in this chapter.

EPICURES' CLUB
Elizabeth, N.J. 07207

CHEESE PACK

A delicious 1-pound wheel of aged Cheddar is the feature of this attractive assortment of fine natural cheeses. Framing the Cheddar are 3-ounce wedges of Bel Paese, Stilton, Port du Salut, Shepherd Girl, rounds of Gouda and Edam, and a 4-ounce loaf of Muenster. $8.50 prepaid.

CHEESE BOX

Eleven varieties of fine cheese to thrill the cheese lover are in this handsome box. Contains hickory-smoked Edam loaf, Kuminost bar, Gouda round, cup of aged Cheddar spread, longhorn Cheddar loaf, Muenster loaf, Edam bar, golden-pure bar, noekkelost bar, cup of blue spread, Edam round. Plenty of good cheese in a cellophane and beribboned package. $10.45 prepaid.

See Meat, Fish and Poultry chapter for company description.

CHEESE 67

FIGI'S, INC.
Marshfield, Wis. 54449

CHEDDAR GIFTS

You have decided to send Cheddar Cheese Holiday Gifts this year. But there's a problem. Betty likes medium Cheddar, but her husband, Tom, likes only sharp. How many couples do you know where that's the case? Sending a wheel of sharp or of medium Cheddar will satisfy only half the family. Now, what kind of a gift is that? Figi's solves your problem by packing two individually wrapped 1½-pound half wheels together: sharp and medium Cheddar in the same box. A perfect family gift! A gift that's sure to please! $7.95 delivered. Also available in one 3-pound wheel of sharp Cheddar for the same price.

ABOUT THIS COMPANY

Figi's is a famous mail-order business which was established over thirty years ago. They specialize in a world of delicacies which include their famous cheeses, smoked meats, sausages, and combinations. And they all come gift-wrapped, for ready giving.

FIGI'S, INC.
Marshfield, Wis. 54449

A VARIETY OF CHEDDARS

Rugged wooden crate contains three 1-pound wheels of Wisconsin's finest Certified Cheddar cheese. Three different ages—three distinct tastes—range from mellow to extra sharp. Yellow—mellow (aged over five months). Red—sharp (aged over ten months). Black—extra sharp (aged over fifteen months). The label on each Cheddar wheel is stamped with the date of manufacture, and is certified to be a premium cheese. $9.95.

See company description in this chapter.

FIGI'S, INC.
Marshfield, Wis. 54449

CHEESE BALL WITH ITS OWN COMPOTE

The height of dainty elegance—a stainless-steel compote arrives to instantly take its place as your snack table centerpiece. Nestled in the charming compote is a giant 16-ounce Swiss Kave Kure cheese ball, rolled in crunchy English walnuts. The five-inch-diameter compote will be used for years as a snack server for spreads, candy, and nuts or will serve as a decorator item to complement any decor. $6.95.

See company description in this chapter.

VERMONT CRAFTS MARKET
Box 17
Putney, Vt. 05346

VERMONT CHEDDAR CHEESE WHEEL

A traditional Vermont breakfast consists of a slice of sharp Cheddar cheese on homemade apple pie. Some folks like to put out a wheel when friends come over for drinks. And it makes a terrific hostess gift. Cheddar is probably the most popular American cheese. Although it originated in the town of Cheddar, Somersetshire, it was the first cheese produced in this country. Usually a hard cheese (sometimes crumbly), the flavor ranges from mild to sharp. Whatever your pleasure, you can't go wrong with a wheel of Cheddar. If you order once, you'll be back for more. $12.95 postpaid.

See Condiments, Spices, and Syrups chapter for company description.

APPLEYARD CORPORATION
Maple Corner
Calais, Vt. 05648

CHEDDAR CHEESE

Did you know that a 1,000-pound wheel of beloved Cheddar cheese was once presented to Thomas Jefferson? From the origins of our nation, it has always been a traditional favorite. And there's nothing like top-quality Vermont Cheddar. Specially supplied through the Cabot Farmers' Cooperative Creamery, this natural Cheddar comes in three delicious variations: sharp, mild, and sage. Each is a tasty addition to your buffet table, Saturday brunch, afternoon picnic, holiday party, or midnight snack. Individually wrapped and sealed for your convenience. Three 8-ounce bars—$6.10. 3-pound wheel—$9.00.

See Condiments, Spices, and Syrups chapter for company description.

NICHOLS GARDEN NURSERY
1190 North Pacific Highway
Albany, Oreg. 97321

CHEESES FROM OREGON'S ROGUE RIVER VALLEY

In southern Oregon where the mighty Rogue River holds dominion and spills into the broad Pacific Ocean, there is a vast area of giant trees and green rolling pastures. In the mid-1800s considerable gold was taken out of this country. When it ran out, the settlers turned to dairy farming, which was the basis of the present-day cheese industry in southern Oregon. They specialize in gourmet-type cheeses which are shipped to all parts of the world under the brand name of "Rogue Gold." Raw Milk Sharp Cheddar (with no color added) is for those who want the natural product . . . not colored, nor heat-pasteurized to kill the beneficial milk bacteria or natural enzymes. Well aged with that good Cheddar tang. 1 pound, $3.95. 2 pounds, $6.25 west of Rockies; $6.75 east of Rockies. Smoked Raw Milk Cheddar is the Raw Milk Cheddar but with the added flavor that comes from being slow-smoked with hardwood alder. For some reason Smoked Raw Cheddar acquires a tantalizing, delicious flavor that is not found in the regular smoked pasteurized cheese. 1 pound, $3.95. Oregon Blue Cheese Wheel is made from an old traditional Danish formula, which gives this cheese a superb, velvety texture with a full, hearty, creamy melt-in-your-mouth flavor. Excellent in salad dressings, delicious eaten along with some crispy crackers, or for making canapés. Time is on the side of this cheese. No one is in a hurry when it is being made. The cheese is allowed to ripen slowly so as to develop it's characteristic blue cheese flavor. Blue cheese is not shipped during the warm summer months. 5-pound Blue Cheese Wheel, $12.98 west of Rockies; $13.98 east of Rockies. In the Salt-Free Cheddar, the same art of cheese making has gone into this cheese as in Nichols' regular Cheddars. Ideal for those on salt-free diets. 1 pound, $3.95.

ABOUT THIS COMPANY

Nichols Garden Nursery has been in business for over a quarter century. They have a most interesting catalog format which is filled with all sorts of unusual herbs and seeds. Send for it—it's free.

PAPRIKAS WEISS IMPORTER
1546 Second Avenue
New York, N.Y. 10028

THE LIPTAUER CHEESE BALL

Here's the big ball of cheese that has everyone saying "Wow!" It practically melts on the knife, spreads like velvet, and sends the taste buds reeling with delight. It's an international potpourri that's an adaptation of an old European recipe, originally made with Liptauer cheese. It takes well-aged American Cheddar cheese, because it's more "spreadable," mixed with a generous helping of sweet Hungarian paprika and Dutch caraway seeds. Shaped in a handy ball, you can carry it with you to picnics, on barbecues, serve it at parties or for family snacks. Spreads on anything. 12-ounce gift box—$3.98 each. 3 boxes—$11.50. 6 boxes—$22.00. 12 boxes—$42.50.

See International Groceries chapter for company description.

MAYTAG DAIRY FARMS
Rural Route 1
Box 806
Newton, Iowa 50208

BLUE CHEESE AND BLUE CHEESE SPREADS

Believe it or not, domestic blue cheese was developed in Iowa in 1936, although imported blue cheese probably originated in France. Commonly confused with Roquefort, the difference is that Roquefort is made from sheep's milk and *only* in

France, while blue cheese is made from cow's milk and is native to other countries as well. Maytag Dairy Farms was selected to produce the new "fancy" blue cheese developed at nearby Iowa State University because of its superior herd of Holstein-Friesian cattle. This herd won more All-American awards during a twenty-year period than any other herd in North America. Maytag blue cheese is a distinctive product with a character all its own. It is rich in vitamins and nutritive value. And it has been ranked superior in flavor to the finest imported cheese. Maytag blue is available in 4-pound ($10.70) and 2-pound ($6.10) wheels. As a spread, it can be ordered in a 20-ounce ($6.30) reusable crock. Ideal for parties, family use, and thoughtful gift-giving. Each package contains a "Guide Book to Cheese Enjoyment," which includes twenty-nine blue cheese recipes.

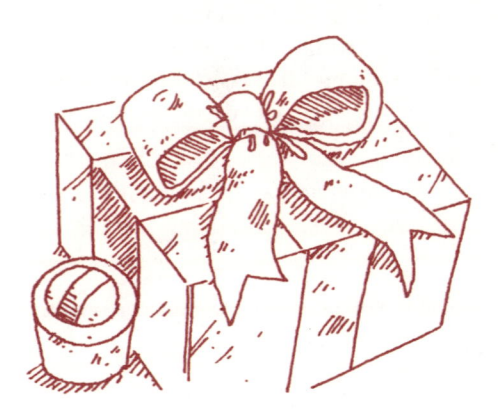

EPICURES' CLUB
Elizabeth, N.J. 07207

CHEESE AND HORS D'OEUVRES

An ideal gift for your friends' holiday cocktail parties because it contains only foods of the finest quality and in generous amounts. Included are smoked turkey pâté, noekkelost cheese, kippered snack spread, boneless and skinless Portuguese sardines, golden-pure cheese, chopped chicken livers, sweet midget pickles, anchovy and olive spread, Edam cheese, Muenster cheese, Norwegian brisling sardines, kuminost cheese, hickory-smoked Edam cheese, and imported cheese from Ireland, Germany, and Switzerland. Sixteen items in all, attractively arranged in a two-part white embossed gift box, cellophane-wrapped and beribboned. $12.45 prepaid.

THE FAMILY PACK

This big, big box of delectables is an ideal family gift because in it there's a thrill for every member of the family. Included are chocolate truffles, Muenster cheese, English rum and butter toffee, butter mints, hickory-smoked Edam cheese, fancy mixed nuts, longhorn Cheddar cheese, strawberry preserves, kuminost cheese, sweet midget pickles, grape jelly, aged Cheddar cheese spread, boneless Holland ham, dry-roasted cashews, stuffed olives, clover honey spread. $17.85 prepaid.

See Meat, Fish and Poultry chapter for company description.

HARRINGTON'S
Richmond, Vt. 05477

AGED CHEDDAR

Here's a real vintage whole milk cheese, cured to the Vermont taste. Cured at the ideal aging temperature of 38°, it's not touched for a year and a half! Only time can produce a fine, sharp Cheddar. There's not much of this kind of cheese around anymore. If you're looking for consistently high quality in the cheese you buy, try Harrington's delicious aged Cheddar. 2 pounds—$5.75. 2½ pounds—$7.65. 5 pounds—$13.25.

See Meat, Fish and Poultry chapter for company description.

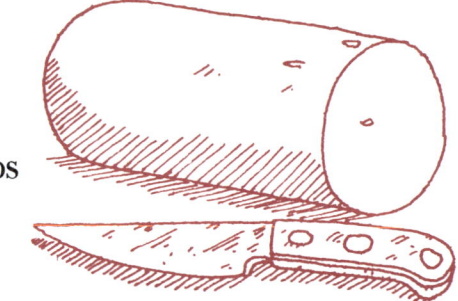

MANGANARO FOODS
488 Ninth Avenue
New York, N.Y. 10018

PROVOLONE

Here's a sharp, tangy cheese that will leave any gourmet in sheer ecstasy! This hard, white provolone, imported from Italy, is perfect, sliced thin, with salami or by itself as an appetizer. Made in a long pear-shaped form, this high-quality cheese is carefully wrapped in foil to preserve its distinctly fresh flavor. Each cheese weighs 1 pound and costs $4.20.

See Special Foods chapter for company description.

SUGARBUSH FARM
Woodstock, Vt. 05091

IN PRAISE OF VERMONT CHEDDAR

Jack Ayres, a 1939 refugee from the New York Stock Exchange turned dairy farmer, became a cheese addict, when one evening he was smoking a piece of ham over an open wood fire. The fire suddenly flared up, touching a piece of cheese on the hearth. Even with that slight smoking, it tasted so good that Jack Ayres, then and there, had an idea on which he started work at once. After many weeks and more experiments, he developed his own process of smoking cheese over hickory and maple fires to impart an incomparable flavor. For three days and nights the smoking now goes on, with Jack checking the golden bars periodically to see if the "look" is right. His Cheddar is aged for many months, much for longer than two years. You'll love its sharp, tangy taste. Choose from a 3-pound medium sharp ($7.95 plus $1.50 postage) or a 6-pound sharp wheel ($13.95 plus $2.35 postage). Either one will leave a distinctly favorable impression on your taste buds.

SWISS CHEESE SHOP
Box 429
Monroe, Wis. 53566

ALPINE BRAND AGED SWISS CHEESE

Here's good news for domestic Swiss cheese lovers who are not satisfied with the slices and bricks available at the local grocery store. Alpine Brand Aged Swiss Cheese is cut from the "heart" of Wisconsin's finest full-cream Swiss cheese, fully aged and selected for its fine flavor and texture. And since it's cut from the center of the cheese, there is no waste. 2-pound square—$5.95.

MERV GRIFFIN'S CARMEL CRAFTSMEN'S CATALOG
P.O. Box 1600
Carmel, Calif. 93921

MONTEREY JACK CHEESE

From where else but the Monterey Peninsula would you expect to find this cheese? It is made from an original recipe by none other than "Monterey Jack," a true-to-life person by the name of Jack, who made this cheese from an old family recipe he brought with him from the Continent almost one hundred years ago. Back in those days when people landed in Monterey, California (mostly whalers and other men of the sea), their first stop for a rare treat was Jack's home for a taste of this mouth-watering delight. At first, the cheese was merely known as "Jack's cheese." Later, Monterey was attached to the name to give a geographical reference as to where someone might find it—giving it its present name, "Monterey Jack." The cheese was always a favorite of sailors as they could take it to sea and not worry about it growing mold for quite a long time, and because of this, the popularity spread to points all along the California coast. Now it's been a hundred years since Jack started making this cheese, and is known the world over for its mild flavor and unusual texture. The original recipe for this cheese is still a secret. We're sure that when you taste this cheese, you'll agree with us in calling it the world's finest and "only" real Monterey Jack cheese.

  Gift-boxed: Half Wheel, 1½ pounds $5.50
       Full Wheel, 3 pounds $7.50

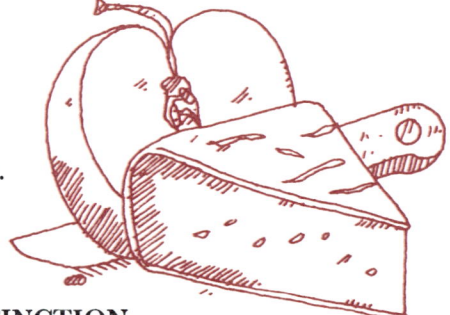

MARIN FRENCH CHEESE CO.
7500 Red Hill Road
P.O. Box 99
Petaluma, Calif. 94952

A CAMEMBERT OF DISTINCTION

Often called the Queen of Cheeses, this world-famous cheese is legendary. Named by Napoleon after the tiny hamlet of Camembert in Orne, France, where it originated, it is mild in flavor, soft creamy to buttery in texture, with a "nutty" tanginess unlike any other soft ripened cheese. Since 1865 four generations of Thompsons working in the family cheese factory on the western slopes of California's Pacific Coast have perfected the art of the Camembert process to the point where gourmets the world over proclaim Rouge et Noir Camembert Cheese the equal of the *Camembert veritable* of Normandie. Rouge et Noir Camembert is best served "straight" as a dessert spread on neutral crackers, with its native goodness complemented by thin slices of apple or pear. But for the adventurous, there are also many recipes to excite the palate. Three 8-ounce portions—$7.18.

BREAKFAST CHEESE

Many are the small, fresh, white cheeses made for that morning meal on the Continent . . . Frühstück, Lunch, Delikat, Lauterbach. Known as the Breakfast Cheese, it is not really spready, but delicately soft, with an old-fashioned ripened butter flavor. Slice it thinly on buttered toast and add a pat of jelly. A fine accompaniment to the song of the lark in the blue and gold morning sky. Cut into cubes, it combines with pineapple, peaches, or grapes to make a fine salad for any time of the day. Six 3-ounce portions—$5.62.

CHEESE

BRIE CHEESE

Popular as a New Year's gift in fifteenth-century France, Brie was described in "An Ode to Brie" by St.-Amant in the seventeenth century as "this gentle jam of Bacchus." Creamy, with an even yellow color throughout, it is at the same time delicate in first flavor with a pleasantly robust aftertaste. The lighter coloring of the surfaces giving way to the darker "reds" is the clue to full ripening and flavor. An eight-inch round of Rouge et Noir Brie on the buffet, flanked by apples, pears, or grapes, is an invitation to a most gratifying taste experience. 24-ounce wheel—$7.09.

SCHLOSS CHEESE

Originated in Austria and named Schlosskäse or Castle Cheese, Schloss was a favorite of the House of Bismarck. Starting along the path of the stodgy Limburger, the process later alters and the finished product takes on the gaiety of its Parisian cousin, Camembert. Rouge et Noir Schloss is a tawny and tangy, yet mellow, little brick, golden in color throughout, with a soft ripened texture. It is a man's cheese, replete with delicate naughtiness, ideal with black pumpernickel and a stein of beer. (If Limburger is the King of Cheeses and Camembert the Queen, then surely Schloss is the Jack of Diamonds.) Six 3-ounce portions—$5.99.

GOLDEN BLOCK CHEESE CO.
Cottage Grove, Wis. 53527

WISCONSIN FULL-CREAM GOLDEN BLOCK CHEESE

If you like your cheese made by the old-fashioned method, you should try this Wisconsin brick cheese. This natural, full-cream brick cheese is made exactly as it was in the early 1900s, when cheese making was at its best. Carefully handmade, one brick at a time, and dry-salt-cured, this deliciously mild cheese is then naturally aged on seasoned wood shelves to bring out its mellow flavor. And the Wisconsin brick cheese is wax-dripped and wrapped before shipment to preserve its high quality. A 3-pound block costs $6.87, and a 6-pound block costs $13.74.

SUGARBUSH FARM
Woodstock, Vt. 05091

BARS OF VERMONT CHEESE

At Sugarbush, they put up their cheese with care. The foot-long bars take time to prepare, but then you get those perfect cracker-sized slices. And the way they put the bars up—each is hand-wrapped in foil (perfect moisture barrier), dipped twice in a special microcrystalline wax which keeps cheese from drying out. Finally, each bar is wrapped in cellophane. That gives you an idea of the care with which Sugarbush tends to their cheeses.

Foot-long Cracker-sized Bars (hickory and maple smoked cheese) $3.65 plus $1.15 postage and handling.

Foot-long Sharp Cheddar Bar $3.65 plus $1.15 postage and handling.

Foot-long Sage Cheese Bar $3.65 plus $1.15 postage and handling.

Foot-long Green Mountain Jack $3.65 plus $1.15 postage and handling.

WIN SCHULER'S, INC.
115 South Eagle Street
Marshall, Mich. 49068

BAR-SCHEEZE

Still serving chips and dips at parties? Are cakes and cookies part of the after-school and pre-bedtime routine? Try something new: Bar-Scheeze comes to you directly from cheese country in Michigan. These tasty spreads are available in three flavors: Tangy Original, Smokey-Bacon with crisp, real bacon bits, and Onion-Garlic, containing the flavor of fresh garlic with bits of sweet Bermuda and green onion. They are a perfect accompaniment to both alcoholic and nonalcoholic beverages, and an ideal, protein-rich, between-meals snack.

18-ounce crock of any flavor	$5.45
16-ounce refills of any flavor	2/$6.95

CONDIMENTS, SPICES, AND SYRUPS

Spices and herbs have played a vital role in man's life for at least fifty thousand years. At one point in history, common salt was used as a base for currency, just as gold is today! Marco Polo's trip to the Far East, as well as the voyages of Columbus and other explorers, were intended to find new ways of bringing exotic, valuable herbs and spices back to Europe.

Even today, there is a mystique that surrounds many spices. Well-known superstition, such as tossing salt over the shoulder, and wearing garlic, have been permanently entrenched in our culture. And in other parts of the world, certain herbs are said to have all kinds of magical, medicinal, supernatural, and religious properties. Although we no longer shed blood over a "bag of saffron," pay our rent in peppercorns, or use bay leaves as protection against lightning, we use these seasonings more than ever.

Besides all of their many other qualities, herbs and spices have one more important feature—they make foods taste better. For example, condiments would never occupy the place they do without interesting seasonings to help them along. In most cases, you couldn't pickle without pickling spices. The variety of condiments and relishes is nearly endless, and thankful we are for that.

And syrups, we mustn't forget about them. What a treat to spread fresh-made pancakes with honey right from the comb, or Grade A Vermont maple syrup, drawn from maples and boiled by special process. What about a wild framboise poured over creamy, homemade ice cream? They are all yours to sample and love. And they make great gifts, too.

INTERNATIONAL SPICE, INC.
6693 North Sidney Place
Milwaukee, Wis. 53209

ADDING SPICE TO YOUR LIFE

Black Pepper Pepper comes from the dried berry (called a peppercorn) of a woody, climbing vine. If the end product is to be black pepper, the berries are pickled while still immature and then dried. The entire berry or peppercorn is used and, as the berry dries, the skin wrinkles and turns black.

White Pepper White pepper comes from the same peppercorn as its counterpart, black pepper; but the berries are left to ripen on the vine for a longer period. The dark skin is then removed and the cores dried in the sun to produce the white peppercorns. These, when ground, give us white pepper. White pepper, although it does not differ significantly in taste, is used where dark particles would be undesirable in the food or sauce being prepared.

Nutmeg Nutmeg, from Indonesia and the West Indies, is the pit or seed of the nutmeg fruit. This sweet-smelling spice is tan in color and is used to flavor a large variety of foods, such as baked goods, puddings, sauces, vegetables, and beverages. The inaccessibility of the source of nutmeg made the spice relatively unknown before A.D. 600. It grows as a peachlike fruit on shiny evergreen trees.

Cloves Imported from Madagascar and Zanzibar, cloves are used in both whole and ground form for studding hams and pork, in pickling fruits, in spicy sweet syrups, and in stews and meat gravies. Cloves are also used in baked goods, desserts, and vegetables. Cloves grow on trees, in bud form. It takes from four thousand to seven thousand dried clove buds to make a pound of spice.

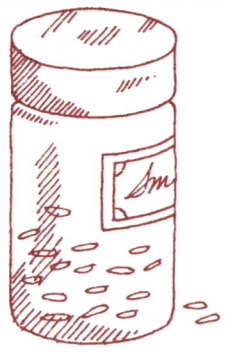

Fennel Fennel was originally considered by Greek wise men to be one of the nine sacred herbs, which could counteract the nine causes of disease. Yellowish-brown in color with an agreeable odor and flavor, these seeds are shaped like miniature watermelons. Most of the fennel seeds are imported from India and Argentina. They are used in breads, rolls, apple pies, Italian sausage, seafood, pork and poultry dishes.

Sea Salt Sea salt is produced by the evaporation of sea water. It contains an abundance of trace materials which make it superior to land-mined salt. Sea salt varies in color from gray to white depending upon the mineral content, processing, and additives which make it noncaking and free-flowing.

All spices come in their own elegant grinder and cost just $3.00 each with a minimum of two. Buy all six, handsomely packaged in a reusable basket, for $10.50.

CONDIMENTS, SPICES, AND SYRUPS 83

WILLIAMS-SONOMA
Mail Order Department
532 Sutter Street
San Francisco, Calif. 94102

FRESH NUTMEG

Like all herbs and spices, nutmeg tastes very much better when it is freshly ground, but nutmeg graters have a nasty habit of grating your fingers as well. We would recommend instead using our comfortable wooden nutmeg grinder, made in Paris. 4″ tall. Comes with a pack of five whole nutmegs and a charming history of this spice, with recipes, by Elizabeth David. $5.00.

See Kitchen Utensils chapter for company description.

WILLIAMS-SONOMA
Mail Order Department
532 Sutter Street
San Francisco, Calif. 94102

IMPORTED HERBS

Evocative of sun-drenched French hillsides, these dried herbs from Provence (tarragon, basil, thyme, and rosemary) are dried on the branch, and consequently the flavors are retained extremely well. To release the oils and obtain the full flavor, it's best to crush herbs in a mortar. Their French maple mortar and pestle is handmade with great attention to the grain of the wood and is particularly attractive to look at as well as comfortable to hold. $7.50 for the set.

See Kitchen Utensils chapter for company description.

MEADOWBROOK HERB GARDEN
Wyoming, R.I. 02898

SPICES

Both herbs and spices are given to us by the plant kingdom, and both share a similar history of service to mankind. Long used for magical, religious, and medical purposes, as well as for their culinary value, spices have added romance and adventure to man's life since ancient times, inspiring poetry, encouraging exploration and trade, and often, alas, bringing rivalry and violence in the wake of greed.

Listed below are some interesting spices. They each cost $.70 per jar and are available whole or ground.

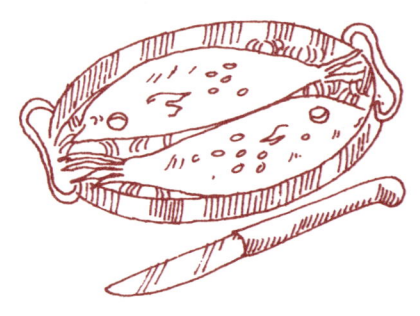

Bay Leaves Makes an unsurpassed addition to soups, stews, fish and meat dishes.

Chili Powder A necessity for most Mexican dishes.

Cinnamon Sticks For stews, sauces, syrups, and hot drinks.

Coriander Try it in pastries, rolls, and cakes.

Fenugreek Used in soups, cheese, stews, chutneys, and curry dishes.

CONDIMENTS, SPICES, AND SYRUPS 85

Juniper Berries Widely used in stews, game dishes, and with sauerkraut. (Ground only.)

Mustard Seed Good with meat, fish, cheese, and egg dishes and mayonnaise.

Pickling Spices A mixture of dill, mustard seed, coriander, and bay leaves. (Whole only.)

Turmeric An essential ingredient of all curry dishes and other international gourmet foods. (Ground only.)

See company description in this chapter.

MEADOWBROOK HERB GARDEN
Wyoming, R.I. 02898

SALAD HERBS

Here's a refreshing combination for a light summer meal: salad herbs. Grown with Meadowbrook's own special biodynamic methods, these delicious herbs are esteemed for their high quality. The mixture includes basil, parsley, celery, marjoram, wild marjoram, dill leaves, tarragon, and chervil—in the perfect combination for your garden salad. And salad herbs can be used in a variety of other ways, too. Try them as a seasoning for poultry, in a butter sauce over carrots, or as a novel way to spark up any food you serve. Each jar of salad herbs costs just $.70.

ABOUT THIS COMPANY
Meadowbrook Herb Garden has celebrated its tenth year in business, specializing in high-quality, low-priced herbs. Write for their free catalog, you'll be enchanted.

WILLIAMS-SONOMA
Mail Order Department
532 Sutter Street
San Francisco, Calif. 94102

SHALLOTS—LILIES OF THE KITCHEN

The French call onions "lilies of the kitchen," and of all the onion family, shallots are undoubtedly the most aristocratic member. They look rather like reddish-brown garlic, sport a pale lavender vest, and are white inside, with a very mild and aromatic onion flavor.

Most shallots are pretty small, which is fine, but taking off the papery skin is time-consuming, so a "colossal" shallot has been hybridized (it's about the size of a boiling onion) for the restaurant trade. Shallots are an indispensable ingredient in French sauces, but chefs don't have time to fiddle about!

These have been packed in 1-pound sacks; use them in place of onions as the flavor is so much better. They keep well in the refrigerator lightly wrapped and of course you can chop them up and freeze them. Colossal shallots, two 1-pound sacks—$5.00.

See Kitchen Utensils chapter for company description.

GNL SHALLOT DISTRIBUTORS
51 DeShibe Terrace
Vineland, N.J. 08360

GOURMET SHALLOTS

This mail-order company began in 1966 and evolved from rather unusual circumstances. Prior to 1966, a group of fellow employees from GNL met on a regular basis to prepare and eat unusual type foods. It was during these sessions that they learned about shallots and particularly their unavailability. Unbeknownst to most people, including the president of the company, was the fact that most of the shallots grown in the United States are grown in southern New Jersey. Not so surprising to this entrepreneur was the fact that his wife's father was one of the South Jersey farmers who raised shallots. The realization that gourmet cooks could not obtain shallots on a regular basis, and that he had an almost unlimited supply available, provided the impetus necessary for a business venture. The philosophy of GNL Shallot Distributors is to supply shallots directly from the farm to the consumer, in any quantity desired, on an *almost* year-round basis. They specialize in subscriptions for ¼ or ½ pound quantities, sent monthly, for a five- or ten-month period. No shallots are shipped in May or June due to inferior quality at that time.

And if you're thinking of an unusual gift to give a gourmet cook, a gift subscription of shallots is a natural. This company features gift subscriptions and includes a gift card with each shipment.

Their prices are as follows:

¼ pound for 5 months	$3.95
¼ pound for 10 months	$7.25
½ pound for 5 months	$5.50
½ pound for 10 months	$9.95

DAKIN FARM
Ferrisburg, Vt. 05456

DAKIN FARM SAMPLER

Twenty-seven thousand cans of Vermont maple syrup and twelve tons of Vermont Cheddar cheese sold from one little country store! This means many satisfied customers. Why don't you join them? Vermont maple syrup needs no description. The name alone recalls the traditional quality of this product. What you may not know about Vermont Cheddar is that it is made from whole (not skimmed) Vermont milk and aged between one and two and a half years.

Try these quality products in Dakin's Farm Sampler: ½ pint maple syrup, a scallop of maple sugar, and a 6" bar of Vermont sharp, smoked, and sage Cheddar. $5.75 to $6.75 postpaid, depending upon where you live.

Write the above address for complete catalog with prices and ordering information.

VERMONT COUNTRY STORE
Weston, Vt. 05161

FAMOUS BERMUDA SHERRY PEPPERS—A DIFFERENT GOURMET CONDIMENT

People who have enjoyed the enchanting island of Bermuda are well acquainted with sherry peppers; a bottle of it stands on every restaurant table and in every good Bermuda home. Everyone always fetches back this rare indigenous treat to use as a special gift which, up to now, has not been obtainable anywhere else. But, happily, for the first time, we can offer it stateside. So what is it? It's a hot, spicy, tantalizing potion made only in Bermuda by steeping peppercorns (and a dozen other spices) in casks of rich sherry wine. It's not to drink! You use it sparingly, drop by drop, to give a rare taste to drinks like bloody marys and foods like fish chowders, soups, and curries. With each bottle, there's a folder of typical recipes for Bermuda foods like kedgeree, buccaneer soup, peppertini, and red hind. This magic elixir comes in a 5-fluid-ounce bottle. $2.50.

NATURAL NEW ENGLAND MINCEMEAT

This traditional rum-flavored New England mincemeat is made from an old recipe that comes from a New England grandmother who made it over one hundred years ago. The tradition has been kept in both kind and quality. This natural, home-style mincemeat is in no way like the commercial type one finds everywhere. A trial in a good home-baked pie will prove the point. It's a big jar—6″ tall, 3½″ wide, holding 1 pound 12 ounces of mincemeat. $1.89.

FIDDLEHEAD FERN GREENS

Once a year a New Englander sends boys out to pick the coiled tender young fronds of this rare and delicate fern. This is a gourmet's delicacy, offered for the first time in tins but known for years as a rare treat by country folk who pick their own. Vermont Country Store fiddlehead fern greens are cooked and need only to be heated and served, or can be drained and served cold in salads. 15-ounce tin—$1.25.

WILLIAMS-SONOMA
Mail Order Department
532 Sutter Street
San Francisco, Calif. 94102

RAOUL GEY IMPORTED VINEGAR

Salad plays a very important part in French cuisine—the three essentials being greens, oil, and vinegar. Fresh young greens can be found, first-pressing olive oil is available if you seek it out, but premium, top-quality wine vinegar is a rare breed indeed. (Less than 1 per cent of all the vinegar made commercially in the United States is suitable for making a first-rate salad.) Raoul Gey wine vinegar is of premium quality, smooth and subtle, with an excellent, deep flavor—not in the least harsh or puckery. The natural white wine vinegar is from champagne, the natural red wine vinegar is from burgundy. 24-fluid-ounce bottles—$7.50 for the pair.

See Kitchen Utensils chapter for company description.

APPLEYARD CORPORATION
Maple Corner
Calais, Vt. 05648

TOMATO CHUTNEY

This tomato chutney is a spicy condiment that gives food an exciting new flavor. Made of fresh citrus fruits, tomatoes, spices, and maple syrup, tomato chutney is skillfully and painstakingly blended in the finest Vermont tradition. Use it for curry dishes, meats, seafood, with cheese and crackers, or with your favorite family recipe. With no additives or artificial preservatives, Mrs. Appleyard's tomato chutney is the perfect way to add new zest to your home-cooked meals. Two 9-ounce jars—$4.70.

ABOUT THIS COMPANY

Ownership of the Appleyards recently changed hands from Edna Kent to two enthusiastic couples who plan on continuing the same fine tradition. The Appleyards has a lovely little catalog from the kitchens of their establishment. And it's free.

VERMONT CRAFTS MARKET
Box 17
Putney, Vt. 05346

VERMONT MAPLE SYRUP

There's nothing finer than the real thing—Grade A Vermont maple syrup. It's made the old-fashioned way by gathering forty gallons of maple sap and boiling it down until it makes just one gallon of pure syrup. Fantastic on pancakes, waffles, and French toast. It also adds an interesting flavor to butter cream icing.

And wait till you try just a little on top of homemade vanilla ice cream! Perfect sugar substitute in cooking, too.

    1-quart tin . . . . $ 7.50 postpaid
    1-gallon tin . . . . $21.95 postpaid

ABOUT THIS COMPANY

Vermont Crafts Market was formed in 1975 to provide you with the finest from Vermont and the New England area. True to their name, their catalog is filled with craft items. Write for their free catalog.

APHRODISIA
28 Carmine Street
New York, N.Y. 10014

CALAMUS ROOT
(*Acorus calamua*)

Sweet flag or sweet sedge, as it is commonly known, has a pleasant taste and aroma. It is often candied and eaten in the same way as angelica. A bit chewed daily is said to destroy the taste for tobacco. In perfumery calamus is used as a fixative, and adds a mellow spicy note to potpourris, especially those with "earthy" or vanilla-like aromas. ¼ pound—$1.95.

CARDAMOM
(*Elettaria Cardamomum*)

Native to India, these aromatic seeds were brought to Scandinavia by the Vikings where they became a necessary ingredient in cooking. They are also used in Swedish meatballs, Danish pastries, pies, marinades, and many curries, as well as in a variety of other ways. Cardamom is a popular flavoring in Indian and Middle Eastern sweets and beverages.

Whole White Cardamom (Middle Eastern, sun-dried) is used in Turkish coffees, Moroccan teas, etc. 1 ounce—$1.50.

Whole Green Cardamom comes from Guatemala and is dried in tumblers. It retains the green color, but is not as aromatic as the white cardamom. 1 ounce—$.90.

Ground Cardamom is used in baking, curries, etc.—such as our banana curry, apple or banana fritters, etc. 1 ounce—$1.10.

Whole Black Cardamom is less aromatic than the other varieties, but is specifically called for in some Indian dishes. 1 ounce—$1.30.

CASSIA BUDS
(*Cinnamomum cassia*)

For years, this exotic spice was not available in the United States because it comes from Mainland China. But with the reopening of trade relations, it is now available. Cassia buds have a tangy sweet taste, much like the combined flavors of cinnamon and cloves. The buds are the dried, unripe fruit of the cassia tree. In the Far East, they are a traditional symbol of good luck and are given to new brides. In cooking, cassia buds are used in confections, Hungarian cherry soup, and sweet pickles. They add a sharp sweetness to potpourris. ¼ pound—$3.10.

DANDELION LEAF
(*Taraxacum officinale*)

The most common of garden pests, it is also known by the names of priest's crown and swine's snout. The name "dandelion" comes from the Latin *dens leonis,* meaning lion's tooth. This is because of the jagged shape of the leaves. Dandelion leaves make a deliciously bitter tea. Many old English recipes call for them as a seasoning for lentils. You can also try them in omelets, soups, breads, jellies, puddings, or as sandwich spreads. Try bread with butter or cream cheese, sprinkled with dandelion leaves, lemon juice, salt, and pepper. As a natural dye, dandelion leaves will turn a fabric a deep magenta. ¼ pound—$1.15.

ROUND-UP PRODUCTS
14 South Barkway Box 135
State College, Pa. 16801

GRIND YOUR OWN FRESH SPICES

If you're a gourmet cook, you know the importance of truly fresh spices. Now you can have the spices you need—with Round-up's fine assortments. These tangy, flavorful spices are chosen for their high quality, then packed immediately for shipment. And because you grind them yourself, they're the freshest spices available. Round-up's spices come in various gift packages, like the Ferdinand Sextet—a combination of whole black pepper, sea salt crystals, whole white pepper, nutmeg chunks, whole cloves, and fennel seed packed in a hand-crafted Madeira gift basket—which sells for $10.50.

WILLIAMS-SONOMA
Mail Order Department
532 Sutter Street
San Francisco, Calif. 94102

A VARIETY OF MUSTARDS

In Paris, Charles Williams discovered a small maison that produces the most delicious, elegantly packed foods Williams-Sonoma has seen in a long time. So many of the old companies have been swallowed up by large corporations, it's always a pleasure to come across a house that still produces foods in the time-honored, time-consuming, old-fashioned way.

Marvelous mustard in 12½-ounce jars . . . the fresh tarragon mustard has fresh tarragon leaves in it and is really delicious; in fact, James Beard took a dollop and pronounced it to be the best he had ever tasted, not at all the usual tarragon paste! Charles Williams' favorite is the lime mustard—a good strong aggressive mustard with definite overtones of lime. $8.00 for the pair.

See Kitchen Utensils chapter for company description.

CONDIMENTS, SPICES, AND SYRUPS

APHRODISIA
28 Carmine Street
New York, N.Y. 10014

CRYSTALLIZED GINGER

Here's a unique, sweet treat: crystallized ginger. This well-known spice is derived from the rhizome of a South Asian plant, then specially prepared to give you a highly appealing food. Try it in baked goods, preserves, chutneys—or as a substitute for your after-dinner mints. However you use it, it's sure to give your foods a spicy, more flavorful taste. One 4-ounce package costs just $.65, and six packages are available at the discount price of $3.70.

Aphrodisia also produces a wide variety of other spices and herbs. For a complete catalog and price list, write the above address.

LORETTE INDUSTRIES, INC.
P.O. Box 27
North Hackensack Station
River Edge, N.J. 07661

"SEASON" WITH BACON, MUSHROOMS, AND GREEN PEPPER

Once you taste these three seasonings (*and smell their aroma*), they'll soon join your salt and pepper as permanent condiments at every breakfast, lunch, and dinner table. These are no run-of-the-mill instant seasonings. Instead, they are artful blendings of natural and artificial flavorings which instantly disperse in any food—hot or cold—to give an unmistakable flavor and aroma. Give your favorite home recipes that "creative touch" with bacon, mushroom, and green pepper seasonings—the perfect way to add new zest to anything you eat. Sprinkle on salads, fish, hors d'oeuvres, soups, pizza, sandwiches, or any other food. And 1 teaspoonful is only 7–8 calories. Three 1½-ounce jars—$6.00.

SUGARBUSH FARM
Woodstock, Vt. 05091

SUGARING SEASON

Sugaring begins in Vermont in mid-March when the days begin to warm up to above freezing. It takes cold nights and warm days to get the sap flowing in the trees. And it takes between thirty-five and forty gallons of maple sap to boil down to one gallon of pure maple syrup! At Sugarbush Farm, they make all their own maple syrup. Sugaring ends in mid to late April, as soon as it warms up, so it doesn't go below freezing at night and the tree buds begin to come out. A limited amount of this grade of maple syrup is made available at the end of the sugaring season. If you would like to order some, it is best for you to order by April, as generally there is not a large supply of this left in the summer and fall months. Most of their customers who have made special arrangements in the past buy their whole year's supply in the spring. Prices are the same, and it comes in the same size cans as their regular Grade A maple syrup. 1 quart—$4.95 plus $1.60 postage and packing. ½ gallon—$9.25 plus $2.50 postage and packing.

DAN JOHNSON
Route 1, Box 265
Jaffrey, N.H. 03452

PURE MAPLE SYRUP

Have you ever tasted the mouth-watering, natural goodness of pure maple syrup? Now you can have nature's sweet richness shipped direct to you from New Hampshire. This luscious, thick maple syrup is made at Dan Johnson's old-time family operation in the foothills of Mt. Monadnock. It's tapped, boiled, and prepared (under rigid state inspection) in the best New Hampshire tradition to give you a thick maple syrup with an unbeatable down-home flavor. Use it for your pancakes, French toast, ice cream, doughnuts, or anything else that needs a sweet touch. However you use it, Dan Johnson's pure maple syrup will brighten up your day!

½ pint	$ 3.30	Six ½ pints	$18.00
1 pint	$ 4.60	Six pints	$26.00
1 quart	$ 6.50	Six quarts	$37.00
½ gallon	$10.50	Six ½ gallons	$59.00

For orders south of Virginia and west of Ohio, add 10 per cent postage.

BREMEN HOUSE, INC.
200 East 86th Street
New York, N.Y. 10028

HONEY FROM AROUND THE WORLD

Honey is one of the most versatile of foods. It can be used as a sugar substitute, as a syrup, or in baking. All of us have long been familiar with the typical American brands. However, you can get gourmet honeys from all over the world. Choose from nineteen varieties (produced in six countries), including acacia, linden, heather, adelshoeve clover, wildflower, black locust, and lotus blossom. Whatever your choice, you're sure to choose a flavor which compliments your taste buds. From Germany: Acacia, $3.50; Linden, $3.00; Heather, $4.00; Wildflower, $2.75. From Holland: Adelshoeve Clover, $3.00. From Italy: Black Locust, $2.50. From New Zealand: Lotus Blossom, $2.50. All come in 16-ounce containers except the Black Locust, which comes in a 13-ounce container. Minimum order: $10.00.

ABOUT THIS COMPANY

Bremen House, located in "Germantown," offers a wide variety of imported German foods and other products. Their catalog is written in both English and German. For a complete catalog and price list, write to them at the above address.

HARRINGTON'S
Richmond, Vt. 05477

HARRINGTON'S MAPLE SYRUP

North country through and through, this Grade A pure maple syrup is tapped from the trees of Vermont. And this state is known *internationally* for its superior syrup. It's hot-packed for long life. 1 quart—$6.85. ½ gallon—$11.00.

See Meat, Fish and Poultry chapter for company description.

EARLY'S HONEY STAND
Rural Route 2, Box 100
Spring Hill, Tenn. 37174

WONDERFUL SYRUPS

Honeypak A honey of a gift. Four 1-pound jars . . . three intriguing flavors: clover, wildflower, and sourwood. (In case of a crop failure, Early's might have to make a substitution but they don't anticipate that.) One jar of clover is all liquid; the other jar of clover has the real honeycomb in it. Clover is a very popular flavor. The wildflower is a mixture of several different honeys—a unique blend. The sourwood is one of the rarest honeys made. It has a distinctive, different, and very delicious flavor. Made from the blossoms of a small tree that grows in the mountains. Included is *My Favorite Honey Recipes* by Ida Kelley—sixty-eight ways to use honey in cooking. All beautifully packaged . . . truly a "sweet" gift. And be sure to order one for your own use. Order "Honeypak" postpaid. Zone A: $6.75. Zone B: $6.95.

Quart Honey is the product that gave Early's Honey Stand its name over fifty years ago. The bees still make it the same way they made it then—pure and unadulterated. Mainly sweet, white Dutch clover. If the apple folks hadn't coined the words "golden delicious," that's what they'd call it! A full quart unbreakable bottle of delicate sweetening. Order "Quart Honey" postpaid. Zone A: $4.15. Zone B: $4.60.

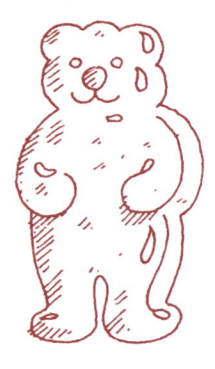

Honey Bear is a cute bear filled with 12 ounces of Early's finest white clover honey. Just squeeze the honey out of the no-drop spout . . . no fuss, no mess. The kids love 'em. Refill and use again and again. The perfect, inexpensive small gift or "thank you" token. Order "Honey Bear" postpaid. Zone A: $2.30. Zone B: $2.50.

Quart Sorghum There aren't many old-time sorghum makers around nowadays, but Early's has one right in their Tennessee hills. They cook it in an open pan just like they did "back then." Beautiful amber color and just the right thickness. They don't add a thing to it. Use it on biscuits, hot cakes—anywhere you want an unusual sweet. Ample quart bottle. Order "Quart Sorghum" postpaid. Zone A: $4.95. Zone B: $5.45.

Little Brown Jugs Everyone who sees these intriguing little jugs falls in love with 'em . . . especially the ladies! Quality ceramic glazed, they are exact replicas of the old-time "coal oil" or "moonshine" jug. Real collector's items . . . and they are filled with delicious pure-bee honey or sorghum—all stoppered and sealed with old-fashioned sealing wax. Sold in pairs only. Unless otherwise specified, we ship one filled with honey and one with sorghum. 12 ounces. Order "Pair of Jugs" postpaid. Zone A: $5.15. Zone B: $5.75.

See Meat, Fish and Poultry chapter for company description.

CONDIMENTS, SPICES, AND SYRUPS

CLARK HILL SUGARY
Canaan, N.H. 03741

PURE NEW ENGLAND MAPLE SYRUP

Add a touch of New England to your table with Clark Hill pure New England maple syrup. It's the syrup neighbors from all over New England come to buy. Why? Maybe it's the way the sweet-water sap is gently lifted from young Indian River maples instead of dripping into buckets. Maybe it's because Clark Hill believes in quality, not quantity, processing the sap with care, one step at a time. Maybe it's because this syrup is as pure as it can be, with no additives or preservatives. Now you can find out for yourself. Wherever you are, you can enjoy New England in this pure maple syrup. ½ gallon at $8.00, quart at $5.00. Also available for your delectation, creamy pure maple butter, the essence of the syrup. 1½ pounds—$5.00. Their old-fashioned pure maple candy is also a treat. 2-pound tin—$9.00.

VERMONT COUNTRY STORE
Weston, Vt. 05161

SLAB OF 1 POUND VERMONT MAPLE FUDGE

Maybe you would like to give (or receive) a 1-pound, 5"×7¾" slab of rich, genuine Vermont maple nut fudge. It's a lot of Vermont maple for the price. $2.50.

MAPLE BUTTER

Maple butter (some call it maple cream) is a spread, made of 100 per cent pure maple, delicious for frostings, toast, cakes, waffles, ice cream, and puddings. Spreads as easily as peanut butter. 12-ounce jar—$2.75.

VERMONT HARD MAPLE SUGAR IN CAKES

Maple cake sugar (hard) can be eaten as confection or melted into syrup, 1 pound of pure Vermont maple sugar cakes. $3.95.

WHITE POPCORN

Unlike the big, tough yellow popcorn sold in movie houses and at carnivals, this popcorn is white, fluffy, tender, and tasty because first, it is clean, with all the impurities removed. Second, it's graded to uniform-size kernels. Third, it's packed in tight bags to retain the right moisture so it will pop. Fourth, it's tender because it pops thirty-eight times the size of the kernel. Keep it in a jar, with holes in the lid, and in the refrigerator to maintain the right amount of moisture. 2-pound bag—$1.75.

VERMONT MAPLE SYRUP

Every year maple prices go up. Farmers can't hire anyone to work, maple orchards are being sold for lumber, and probably, if the farmer included cost of equipment and his time, maple syrup would cost twenty dollars a half gallon! You can buy maple syrup, not guaranteed, on some roadside stands for much less money, but you will not be getting the same quality. The Vermont Country Store guarantees their maple syrup. That means that you get exactly eleven pounds to the gallon, because if it is less, it will ferment, and if it is more, it will precipitate hard sugar. It's important to buy maple syrup from people who guarantee it.

```
1-pint tin ................................. $3.25
1-quart tin................................. $5.75
½-gallon tin................................ $8.95
```

HARRY AND DAVID
Medford, Oreg. 97501

HOMEMADE RELISH ASSORTMENT

Keep an unusual assortment of relishes on your shelf that will really do justice to the meals you serve. There's Bear Creek relish—sweet-and-sour tangy with lots of Hawaiian pineapple, fresh cucumbers, and spices. Wonderful with Thanksgiving turkey, Christmas ham, hot dogs, hamburgers, steaks. Cherrydills are sweet-

CONDIMENTS, SPICES, AND SYRUPS

pickled Royal Anne cherries—something Grandma never made, because it's a new recipe. Scrumptious with pork, poultry, salads. There's an unusual sweet pickled watermelon rind—put up the old-fashioned way, with lots of fresh spices and a cinnamon stick. And an all-purpose meat sauce—a woodsy, smoky blend made with ripe tomatoes, pure honey, onions, cider vinegar, and savory spices. So good, it won the Better Homes and Gardens outdoor cooking contest and the Gold Medal at the California State Fair. All four (2½ pounds in all) are put up in authentic old-fashioned heavy glass jars that you'll use all year for spices, candy, condiments. $11.95 delivered.

See Confections chapter for company description.

GOKEYS
21 West Fifth Street
St. Paul, Minn. 55102

THE SIOUX AND ALGONQUIN FOUGHT WARS FOR 250 YEARS TO GET WILD RICE

It's a true story. The Sioux were plains Indians, and the Algonquin were river Indians. Their life styles were markedly different. But so much did they both prize the wild rice territory in Minnesota and Wisconsin that they fought repeated battles over it. Wild rice was called "Mahnomen" by the Indians, literally "good berry." And it was a perfect description. In addition to possessing a most unique flavor and quality, wild rice is also more nutritious than ordinary rice. Its protein content, for example, is 14.1 per cent versus 6.7 per cent for white rice. No wild-game dinner is complete without it. Delicious with just melted butter and salt and pepper or prepare it in casserole or in dressings. You will have to agree that the delicious, rich flavor of natural wild rice is unsurpassed. Complete cooking instructions included in each bag. 1-pound bag—$6.50.

VERMONT COUNTRY STORE
Weston, Vt. 05161

MEALS AND FLOURS

Listed below are nine different stone-ground flours and meals for baking everything from bread to cake.

Bread Wheat Flour, sometimes called "graham flour" because years ago **Dr. Graham** advocated it for good health. This is the real germ-content high-protein flour. The best dark bread and rolls can be made from this rich flour without using any white flour.

Pastry Wheat Flour is a whole-grain stone-ground flour for making delicious pastries such as pie crust, tarts, cakes, and cookies. This flour contains less protein than bread flour and that's why some like it for pastries and the like.

Yellow Corn Meal There are over three hundred ways to use corn meal. This yellow corn meal is rich and full-flavored and preferred by folks up north.

White Corn Meal The sweet fine white corn meal is wonderful for southern dishes, and is preferred by folks south of the Mason-Dixon line and in Rhode Island.

Rye Flour Old-fashioned stone-ground rye meal for bread and other uses.

Oat Flour A ground oatmeal flour for use in making oatmeal bread, cookies, etc.

Muffin or Pancake Meals By grinding together whole-grain corn, wheat, and rye, this contains the good taste and the vitamins of America's three most nourishing grains! Easy to use for pancakes, muffins, rolls, etc.

Buckwheat This 100 per cent pure buckwheat flour (not a mixture) contains only buckwheat.

Rich High-Protein Soy Flour The richest vegetable source of vital protein, just a little added to other whole grains gives you a zestful source of protein. Use one tablespoon of soy flour for each cupful of their other whole grains.

5-pound bag—$2.50.

CONFECTIONS

If you believe that no meal is complete without a fancy dessert, or a delicious, sweet delicacy, then you'll be delighted with the tempting morsels described in this chapter. Do you dream about hand-dipped chocolates, or fresh-made pecan pralines, or delectable marzipans, or sumptuous tortes, tarts, cookies, cakes, and more? Well, here they are.

Feast yourself with an endless variety of confections. Satisfy your craving for fresh, chocolate-covered cherries; savor the piquant tang of Malaysian pineapple; enjoy the special chewiness of honey-dipped dried fruits; partake in the ambrosia of a brandied fruitcake that's rich enough to drink. Treat yourself to a *real* pie that's made with fresh, mammoth pecan halves. Serve roasted and fresh nuts that are an epicurean inspiration. Try a candy confection made from carob powder and blackstrap molasses. Let your mouth water over the soft center chocolate pieces which have been made in family kitchens.

And if the aroma of Grandma's jams and jellies still frequents your thoughts, you can re-create the old days with a wonderful selection of jams, jellies, preserves, and marmalades that will put you right back in Grandma's kitchen. Now you can have that same high quality, whether you select ginger preserves, strawberry marmalade, or more traditional fruit flavors. You'll love their richness on brioche, toast, or as a quick-energy treat.

Elaborate dinners are made even more exciting when they include a Spitzkuchen (honey cake with chocolate) and ginger johnnies—even an edible ginger house! Indulge yourself. All these spectacular treats are only as far away as your mailbox.

BUTTERFIELD FARMS, INC.
8500 Wilshire Boulevard, Suite 1005
Beverly Hills, Calif. 90211

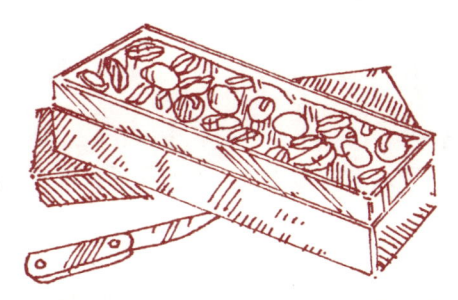

GOURMET FRUITCAKE

Flavorful French cherries, Giant Charbert walnuts, Malaysian juice-rich pineapple, plump, meaty, Spanish Marconna almonds, domestic midget pecans, and California wine raisins, with just enough cake to hold them together! Doesn't that sound like a gourmet fruitcake? Maybe for some, but not for Butterfield. Once baked, this rich cake is slowly aged in a combination of 100-proof bonded Kentucky Bourbon, 86-proof New England rum, and 80-proof brandy. The longer it's aged, the better it tastes. An ideal way to say thank you to relatives and friends, a treat for the whole family, Butterfield's is the gourmet fruitcake that's "good enough to drink."

Available in 1-pound, 2-pound, and 5-pound gift packages at $5.95, $8.95, and $15.95 respectively, plus shipping. Contact Alvin Feidel at the above address for complete ordering information.

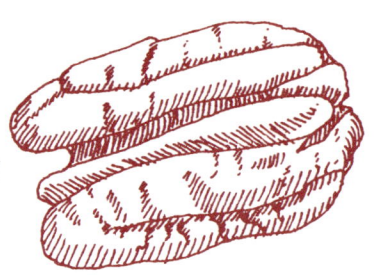

STERNBERG PECAN COMPANY
Dept. D
P.O. Box 193
Jackson, Miss. 39205

STERNBERG'S WHOLE PECAN HALVES

Remember eating pecans you shelled with a nutcracker? Now you can have freshly shelled pecans without any of the work. These delectable pecans are available from Sternberg, a concern unique in the pecan business because pecans are their only product. They offer fancy-grade mammoth shelled whole pecan halves. When a Sternberg pecan reaches the packing box, it has gone through every step of grading and sorting known to the industry. An additional step makes Sternberg pecans worth buying: the final, old-fashioned, personal inspection by someone who cares. A recipe folder including the old, old method of preparing pecan pie is enclosed in every package. Available in packages of 2 to 10 pounds, Sternberg's mammoth whole pecan halves make an excellent and unique gift for yourself and for gourmet friends.

2 pounds	$ 7.50
3 pounds	$10.75
5 pounds	$16.50
10 pounds	$30.00

ABOUT THIS COMPANY

Sternberg's, founded in 1938, is a family concern which specializes in selling pecans via mail order. They are located in the heart of the pecan-growing district, which supplies the entire world with these pecans.

SUNNYLAND FARMS, INC.
Albany, Ga. 31702

BLACK WALNUT PIECES

Fancy large-size black walnut pieces from the Ozark Mountain country. Grand for dessert toppings, sprinkled on salads, for cakes and cookies. The pungent black walnut aroma adds zest to most recipes that call for "nuts." (Black wal-

CONFECTIONS

nuts are not to be confused with English walnuts, which are made more bland.) Select large pieces only. Black Walnut Fact Folder and recipes in each box.

    2-pound 13-ounce Home Box $10.15
    5-pound Home Box $16.30

See company description in this chapter.

SUNNYLAND FARMS, INC.
Albany, Ga. 31702

ROYAL MIX (ROASTED AND SALTED)

A party nut mixture much better than you can find anywhere else. More than a quarter of the total mix is Sunnyland toasted mammoth pecan halves (whoever heard of so many pecans in mixed nuts?). Another 15 per cent is hickory-smoked almonds, which adds an unusual zest. The rest are top-quality Brazil nuts, cashews, and filberts fresh-roasted in their own kitchens.

    1-pound 12-ounce Mixed Nuts Gift Tin .... $ 7.75
    4-pound Mixed Nuts Home Box $13.10

See company description in this chapter.

SUNNYLAND FARMS, INC.
Albany, Ga. 31702

CHOCOLATIER

Chocolate lovers, *Attention!* Here are three of the best chocolate pecan delights you can get. Toasted pecan halves covered with rich chocolate (Choco-Nuts); rich creamy home-style fudge filled with small pecan pieces; and the light and airy Pecan Divinity completes the treat. 2 pounds 9 ounces—$9.90.

See company description in this chapter.

SUNNYLAND FARMS, INC.
Albany, Ga. 31702

CAROB PECAN BARK

A new candy especially for people who enjoy a sweet but prefer to avoid chocolate. This chocolatey bark blends carob powder, blackstrap molasses, wheat germ, rose hips, and lots of pecans. Carob trees grow around the shores of the Mediterranean Sea. They provided the "locusts" which John the Baptist lived on (along with honey) in the wilderness. When dry, carob beans rattling in their pods sound just like locusts. 1 pound 11 ounces—$6.90.

See company description in this chapter.

SUNNYLAND FARMS, INC.
Albany, Ga. 31702

PECAN CAR-A-MELLOS

If you go nuts over sweets, here's one you shouldn't miss: Pecan Car-A-Mellos. These delicious little morsels are made of marshmallows dipped in thick, creamy caramel, then rolled in fresh pieces of luscious pecans. The result is a mouth-watering treat that's sure to be a hit anytime. Serve them to guests after gourmet dinners, pack them in children's lunch boxes, keep them handy for your afternoon teatime. Eat them once—and your taste buds will keep craving for more. A gift box of 15 is only $6.50, or you can order 4 boxes for only $21.00.

ABOUT THIS COMPANY

Since 1926, the Willsons have grown fine pecans on their own grove in the heart of the pecan-growing country. Send for their free, 32-page, full-color catalog.

BAILEY'S OF BOSTON, INC.
26 Temple Place
Boston, Mass. 02111

CHOCOLATE LOVERS' DELIGHT

Can you imagine the luscious flavor of a handmade chocolate fresh from the kitchens of one of America's most experienced candymakers? If you're a chocolate freak, you'll love Bailey's soft-center dark and milk chocolate pieces. Write for their famous assorted chocolates, hand-dipped soft-cream centers, fruits, jellies, nuts, and chewies. Available in 1-, 2-, 3-, and 5-pound packages, they cost just $3.25 per pound.

See company description in this chapter.

BAILEY'S OF BOSTON, INC.
26 Temple Place
Boston, Mass. 02111

FRESH-ROASTED NUTS

If you love the taste of all kinds of nuts, then you'll be glad to learn about Bailey's of Boston. Famous old candymakers, they also sell delectable, fresh-roasted nuts with a real quality taste—something that's becoming more and more difficult to find. Choose from peanuts, almonds, pecans, cashews, pistachios, nutchews, and many others—or buy an assortment of selected mixed nuts. All are guaranteed to give you that delicious, fresh, nutty taste you crave. Salted Spanish peanuts—$1.35. Salted almonds (fall season only)—$1.90. Salted pecans (fall season only)—$2.50. Salted cashews—$4.25. Red pistachios—$3.80. Nutchews (almond, cashew, pecan, peanut)—$3.50.

See company description in this chapter.

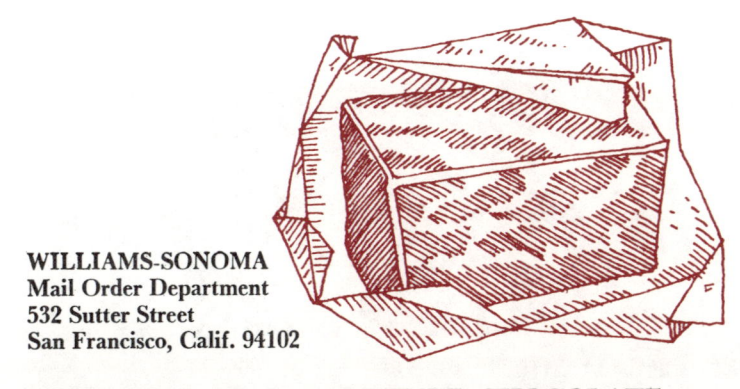

WILLIAMS-SONOMA
Mail Order Department
532 Sutter Street
San Francisco, Calif. 94102

10 POUNDS OF COUVERTURE CHOCOLATE

Van Leer's semisweet chocolate is manufactured for discriminating professional candymakers and pastry chefs. It comes in eye-widening 10-pound bars (that's over a foot long and nine inches wide!) and it smells divine. It also tastes divine

—with any luck it will arrive already cracked, courtesy of the post office, and you can blissfully nibble on the shards in the name of tidiness—and it just happens to be a fantastic cooking chocolate.

It's perfect for coating cakes and petits fours, making shaved chocolate curls, hand-dipping candies, and putting into all kinds of cakes and desserts. It keeps well, in reasonably cool surroundings, *if* you can keep the family away from it. $25.00.

See Kitchen Utensils chapter for company description.

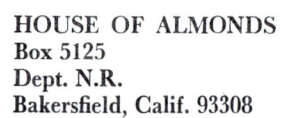

HOUSE OF ALMONDS
Box 5125
Dept. N.R.
Bakersfield, Calif. 93308

PANTRY JAR ALMONDS

If you love almonds, here's the ideal way to keep them around: Pantry Jar almonds. These tasty little nuts are sliced, diced, and slivered six different ways, then packed in old-fashioned glass pantry jars to enhance your kitchen decor. Use them in soufflés, vegetable casseroles, noodles and tuna, baked haddock, garden salads, or even just by themselves. They're a handy means of giving your favorite recipes new zest and flavor. Six ½-pound jars cost $7.95 delivered.

ABOUT THIS COMPANY

The House of Almonds has an 18-page, full-color catalog that's chock full of almonds, dates, and other special confections. It's free for the asking.

BAILEY'S OF BOSTON, INC.
26 Temple Place
Boston, Mass. 02111

OLD-FASHIONED CREAM MINTS

Delight your taste buds with the cream mints you dream about. These handmade concoctions are perfect as an after-dinner treat, with your afternoon tea, or for pure, unadulterated enjoyment! They are available from Bailey's, one of

CONFECTIONS

America's most experienced candymakers. And they come in a lovely variety of flavors: peppermint, wintergreen, lemon, lime, orange, and clove. They are $2.25 a sampler.

ABOUT THIS COMPANY

Fine chocolates are a tradition at Bailey's of Boston, which has been producing quality confections for over one hundred years. You may want to sample their many candy treats—including a rich variety of chocolates, hard candies, sauces, and nuts. For a flier and price list, write to them at the above address.

SUGARDALE FOODS, INC.
1600 Harmont Avenue N.E.
P.O. Box 8440
Canton, Ohio 44711

SUPER PACIFIER

Here's a unique gift idea for the big baby in your family—a Super Pacifier. This 11-ounce milk chocolate treat *is shaped in the form of a baby's pacifier*. It's the perfect way to soothe hurt feelings or calm anxious nerves. Give it to harried executives, high-strung secretaries, nervous relatives, or anyone else who needs a soothing sedative. You and your friends will get a kick (and a lick) out of this novelty "security blanket." Cost is just $5.75, including postage and handling.

APPLEYARD CORPORATION
Maple Corner
Calais, Vt. 05648

APPLE CIDER JELLY

Do you love the taste of cider on a warm autumn evening? Then you'll love Mrs. Appleyard's apple cider jelly anytime. Made of Vermont's best and most delicious apples, this sweet, yet tangy jelly is great with toast, hot muffins, crackers, biscuits, bagels, pancakes, and just plain bread. And because it contains no sugar, artificial sweeteners, additives, or chemical preservatives, this apple cider jelly is the natural way to treat yourself to an old-fashioned favorite. Two 9-ounce jars —$4.70.

See Condiments, Spices, and Syrups chapter for company description.

SUGARDALE FOODS, INC.
1600 Harmont Avenue N.E.
P.O. Box 8440
Canton, Ohio 44711

PRETZEL JOYS

Here's the delicious combination of pretzels and chocolate in a new taste treat: Pretzel Joys. Bits of pretzel and chocolate are skillfully blended together, then mixed in a sugary white mound to make a delicious candy. Price is just $3.50 for a 13½-ounce can.

KAKAWATEEZ LIMITED
130 Olive Street
Findlay, Ohio 45840

TOTEM MACADAMIAS

Looking for a new nutty treat? Try the exotic taste of Totem macadamias. These delicious gourmet nuts are specially grown in Mexico and are renowned for their distinctly different flavor. Delicately prepared through Kakawateez' own unique dry-roasted process, Totem macadamias are guaranteed fresh and crisp. They're just right for snacks, hors d'oeuvres, or even your favorite dessert recipes. Taste them once—and you may never be satisfied with plain peanuts again. A 6½-ounce jar costs just $2.59.

COLONIAL GARDENS
270 West Merrick Road
Valley Stream, N.Y. 11582

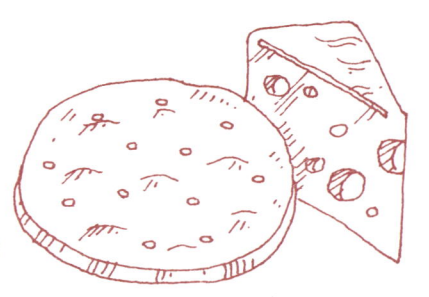

CLASSIC TABLE WATER CRACKERS

Now you can nibble in elegance—with delectable table water crackers imported from England. These delicious edibles go great with cheese, soups, preserves, hors d'oeuvres, or just by themselves. They're perfect for that afternoon tea or late-night snack. They are made by Carr's—a tradition of England—which also has a variety of other crackers and biscuits—all equally tasty. You can have your biscuits and crackers sent in one of over twenty-five differently designed tins—including some decorated with famous English landmarks. 12-ounce tin—$4.59.

See Kitchen Utensils chapter for company description.

CONFECTIONS   111

WONDER BAR PRODUCTS
2323 Haviland Avenue
Bronx, N.Y. 10462

GOUR MATE

This gourmet's delight makes a perfect gift for any occasion: birthdays, anniversaries, graduations, housewarmings, and holiday gift lists. An imported Zagreb ham, Jasmine tea, Scottish fruit preserves, an assortment of Scandinavian cheese spreads, and Strawberry Candies, all packed in a bamboo basket that can double as a fruit bowl. A selection to satisfy everyone. $9.95 delivered.

WILLIAMS-SONOMA
Mail Order Department
532 Sutter Street
San Francisco, Calif. 94102

DELECTABLE IMPORTED PRESERVES

The selection of Raoul Gey preserves includes black cherry, a miraculous affair of closely packed, small whole pitted cherries suspended in cherry jelly; mirabelle plum, a wonderful concoction of yellow, sun-ripened plums from Alsace very dear to the French; and preserved orange slices, which consists of closely packed, tender slices of whole orange in an excellent bittersweet preserve. These are all delicious with croissants and toast, of course, but they are *absolutely outstanding* with ice cream. Set of three 15¾-ounce jars—$11.00.

See Kitchen Utensils chapter for company description.

COLONIAL GARDENS
270 West Merrick Road
Valley Stream, N.Y. 11582

TIPTREE PRESERVES AND MARMALADES

From Wilkin and Sons, Ltd., of England come these fine, flavorful preserves and marmalades. Each is made from freshly gathered fruits grown on the lush, lavish orchards and gardens of Essex County. Whether you prefer logenberry preserves, ginger preserves, lemon marmalade, or strawberry marmalade, you'll be delighted with these delicious preserves and marmalades. And they're all hand-cultivated, hand-preserved, and hand-packed—to give a unique natural flavor. Available in 12-ounce size.

    Logenberry Preserves $2.49
    Ginger Preserves $2.39
    Lemon Marmalade $2.19
    Scarlet Strawberry Marmalade $3.19

See Kitchen Utensils chapter for company description.

HOUSE OF WEBSTER
Box 488
U.S. Highway 62 North
Rogers, Ark. 72756

STRAWBERRY AND PEACH PRESERVES

Can you imagine receiving a preserve that is so fresh that it is *still warm* when shipped? These luscious sweets are made of the finest, freshly picked fruits available—then prepared according to Webster's own superb recipe to bring out their rich, juicy flavor. They're a delectable way to complement your muffins, bagels, biscuits, crackers, waffles, even bread. And each preserve comes in its own delightful reusable container. Choose from peach preserves in a 1-quart ceramic cream can; 2 pounds 8 ounces of strawberry preserves in a ceramic strawberry; or 1 quart of strawberry preserves in a ceramic pot. Each is a delicious addition to your decor, and each costs $7.75 delivered.

GREAT VALLEY MILLS
Route 309
Quakertown, Pa. 18951

CHRISTMAS STOLLEN

From the heart of Pennsylvania Dutch country comes the legendary German coffee cake, Stollen. It is marvelous for a Christmas breakfast or an evening buffet. It is shaped in a long loaf, baked with Great Valley Mills unbleached flour, and loaded with citrus fruits and raisins washed in butter. Immediately

CONFECTIONS 113

after it is baked, the Stollen is sealed in foil to ensure its freshness and flavor. It's a real treat for the holidays. 1½ lbs.—$6.50 delivered. (No shipments of Stollen before November 15.)

Serve this marvelous Christmas Stollen on a ceramic bread platter. This 7" ×13" hand-painted, wheat-patterned bread platter is one that can be passed along as a family heirloom. Truly an exquisite piece to own or give. $8.95 delivered.

The Great Valley Mills also offers an assortment of Pennsylvania Dutch smoked meats: turkey, ham, sausage. These specially cured delights are particularly great for the holidays—but can be eaten with pleasure any time of the year. And all meats are fully cooked, so you can eat them immediately after they arrive.

See Special Foods chapter for company description.

COLLIN STREET BAKERY
Corsicana, Tex. 75110

THE ORIGINAL DELUXE FRUITCAKE

You may like fruitcake, but have you tasted Deluxe, the fruitcake preferred by European royalty and the American elite in every field of endeavor? From a recipe originated by August Weidman in Wiesbaden, Germany, in 1895, Deluxe has titillated palates in America and abroad for eighty years. More than 75 per cent of Deluxe customers—some going back three generations—place orders year after year. Why? Deluxe contains less than 20 per cent batter. The remaining 80+ per cent consists of candied Oregon cherries, Hawaiian pineapple, sultanas, citron, Italian orange peel, and 27 per cent new-crop pecans from Georgia and Texas. It is hand-decorated with additional cherries, pineapple, and pecans, and glazed with Texas cotton blossom honey. Each cake is baked to order and mailed to arrive on the date you specify. Family, friends, and business associates deserve this special treat. 2 pounds—$5.95. 3 pounds—$8.35. 5 pounds—$13.45.

ABOUT THIS COMPANY

The Collin Street Bakery has quite a legend. Their first big order was for the troupers of John Ringling's circus. An early-day customer roster included Enrico Caruso, Will Rogers, John J. McGraw, and "Gentleman Jim" Corbett. Awards include that of the New York Gourmet Society and Will Rogers, Jr. A booklet in every cake tin tells the cake's history, how to slice and preserve.

BYRD COOKIE COMPANY
P.O. Box 13086
Savannah, Ga. 31406

WILD RICE TEA CAKES

Indian folklore has it that wild rice must be harvested from the shallow lakes and rivers in the Northern Middle Lakes States. The last week in August and the first two weeks in September is when Wild Rice is harvested. To properly harvest a pad of wild rice, so the folklore goes, you must go in three or four times and beat off the wild rice with a stick into the bottom of canoes. Now you can savor this folklore in Byrd's Wild Rice Tea Cakes. Made from selected wild rice kernels, these very unusual cakes can be served alone or spread with butter or fruited cream cheese spreads. Packed in a 7-ounce metal, airtight tin, they cost just $3.90.

BYRD COOKIE COMPANY
P.O. Box 13086
Savannah, Ga. 31406

BENNE BITS

When slaves first came to the coastal areas of Georgia and South Carolina, they brought with them—as their most valued possession—a little handful of benne seed which they believed held for them the secret of health and good luck. Planted near the slave quarters of the early plantations, benne became a traditional part of the Old South. Cooks in the "big house" kitchens knew just how to use this rich spicy, honey-colored seed to make candy, bread, etc. Today, these sesame-like seeds are combined with cheese, flour, and spices to give you a new taste treat to serve with your favorite beverages. Subtle and appetizing, these bite-size crackers are packed in a 9-ounce metal, airtight tin. $3.90.

H. M. THAMES PECAN CO., INC.
P.O. Box 2206
Mobile, Ala. 36601

AZALEA BRAND PECANS

End your holiday shopping worries! Send something different that will delight both friends and business associates: jumbo pecan halves, the pick of the South's finest crop. Crisply delicious by themselves or roasted in butter for chewy hors

CONFECTIONS

d'oeuvres. Great in fruitcakes, pies, dressings for the holiday birds, candy, and pastries. Thames chooses only the finest pecans to shell for you, inspects them dozens of times throughout the shelling process, and then selects only the best to mail to you. Satisfaction guaranteed or money back within five days of receipt. Gift packages from $7.50.

PECAN COOKERY FOR THE GOURMET

A special surprise to enclose with your gift of pecans: this informative and creative gourmet cookbook. Forty years of quality recipes from satisfied customers in 116 pages. Appetizers, breads, cakes, candies, cookies, salads, vegetables, and main dishes. Imagine preparing Tipsy Chicken with Pecans marinated in white wine! . . . or baked stuffed fish with delicate Pecan-Grape Sauce! And nearly forty different recipes for pecan pie!! Only $2.00 ($1.50 when sent with a gift package of pecans).

Write the above address for catalog and ordering information.

BISSINGER'S
205 West Fourth Street
Cincinnati, Ohio 45202

LIQUOR-FLAVORED HARD CANDIES

On the wagon? Or if you are one of those people who drink for the flavor rather than the effect, here is a candy for you. Now enjoy the smoky flavor of scotch, the dry, juniper berry tang of gin, and the deep, rich taste of Bourbon in hard sucking candies. No alcohol, just flavor. Order single flavors packed in 12-ounce brandy snifters ($2.95), or a combination of all three flavors in a sectional glass "Liquor Tower" ($5.95). Keep them around the house, in the office, or offer them as unique gifts. All prices plus postage and handling.

Write the above address for a complete catalog and ordering information.

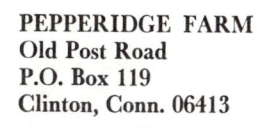

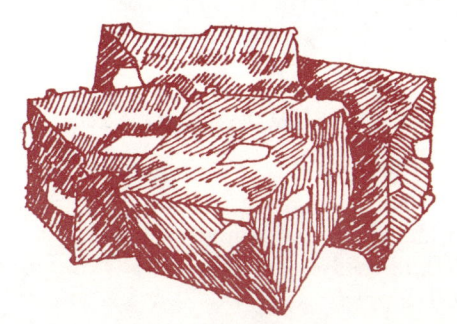

PEPPERIDGE FARM
Old Post Road
P.O. Box 119
Clinton, Conn. 06413

YUMMY FUDGE

This old-time favorite, rich and creamy, recalls the anticipation we felt when it stirred on the stove at home. This Fudge Duo, from Pepperidge Farm, includes plain old-fashioned fudge plus walnut fudge—in their own individual pans. All pure and natural ingredients. 12 ounces of each fudge—$6.95.

FRUIT SPREADS

Full-flavored fruits, simmered at low temperature to preserve the natural taste, make a wonderful fruit spread. Not as sweet as ordinary preserves—and so good for covering a slice of bread. Or try some on ice cream and other desserts. This is a treat everyone enjoys—one of their most popular choices. All natural ingredients, no preservatives. Three jars, 10 ounces each—$6.95.

ABOUT THIS COMPANY

We all remember Pepperidge Farm. It's a name which has been around for years in our local stores. Always famous for their baked goods, they have now expanded to gourmet foods available through the mail only. Send for their free gift catalog.

WILLIAM PENN GIFTS
6515 Castor Avenue
Philadelphia, Pa. 19149

CASTAGNE BASKET OF SPECIAL TREATS

Large walnut-stained Chestnut Wood Castagne Basket from Italy has rope handles, leather trim. Filled with Ferrara Rhum Cakes, Danish Butter Cookies, Koopers Rum Cordials, Jacobs English Cheese Biscuits, Italian Hazelnut Chocolates, Romanoff Cocktail Pâté, French Pure-Fruit Drops, Mr. Gona Cheese, Mint Peps. Exotically topped off with folding fan and hot pink/orange ribbons. $33.95 delivered.

See Kitchen Utensils chapter for company description.

CONFECTIONS

WILLIAMS-SONOMA
Mail Order Department
532 Sutter Street
San Francisco, Calif. 94102

LIQUID CARAMEL AND POWDERED VANILLA

Making caramel is a tricky operation—it tends to turn solid and stick to the pan amid nasty burning smells. Liquid caramel from France is accented with pure vanilla and has a lovely flavor—good for custards and ice creams. You can also use it for foolproof orange glacées and for brushing over apple tarts just before serving. The same company also makes powdered vanilla, which is pure dried and ground vanilla bean. Use it in any recipe that calls for vanilla. Two 4.5-ounce bottles of liquid caramel and one 7-ounce bottle of powdered vanilla—$6.25 for the set.

See Kitchen Utensils chapter for company description.

MISSION PAK
Santa Clara, Calif. 95050

AND VISIONS OF SUGARPLUMS . . .

Memories of an old-fashioned Christmas . . . the whole family gathered for fun and feasting . . . the scent of roast turkey drifting from the kitchen . . . the stuffing and mashed potatoes already steaming on the festive candlelit table. And before that meal of meals was over, everybody in the room was reaching for the most treasured tradition of all—the big bowl of sugarplums. You can bring back memories like this with Mission Pak's indescribably luscious sugarplums. Plump, full-flavored, dipped in chocolate—a delight to old and young alike. The pitted plums are moist and tender, the chocolate dark and creamy-rich. Children who've never tasted sugarplums before will be thrilled to discover that they are actually "for real"! Why not give sumptuous sugarplums this year . . . and remember to give yourself some too! Available year round.

  1-pound box . . . . $6.95 delivered
  2-pound box . . . . $9.95 delivered

MISSION PAK
Santa Clara, Calif. 95050

HAWAIIAN HEAVEN

Macadamia nut cake at its moistest tantalizing best! The rare nuts come from the gentle slopes of Hawaii's famed Mauna Loa volcano where growing conditions are perfect the year round. Prepared in a rich butter batter from an honored, age-old Polynesian recipe, this legendary delicacy is filled with generous chunks of flavorful pineapple and studded with meaty macadamias throughout. There is nothing more Hawaiian than Diamond Head, grass skirts, and Mission Pak's heavenly macadamia nut cake. Send a priceless piece of the Islands to some favorite sweet tooth! Available all year round.

> One 1¾-pound cake .... $ 5.95 delivered
> Three cakes (to one address) .... $13.95 delivered

MISSION PAK
Santa Clara, Calif. 95050

FRUIT-SHAPED MARZIPAN

Marzipan candies have long been a holiday flavor tradition in Europe—and they're fast becoming one here, too. These old-fashioned delicacies stir nostalgic memories in grownups, while youngsters are discovering a "new-fashioned" taste sensation that's out of this world! Almond-flavored, the colorful candies are shaped like miniature apples, pears, plums, oranges, and strawberries—and they're every bit as bright and welcome as the fruits they resemble. Beautiful to look at, they're even better to eat! Once you've tasted them, you'll find yourself coming back for more. A big 11-ounce treat for every sweet tooth in the house. Available the year round. $5.95 delivered.

MISSION PAK
Santa Clara, Calif. 95050

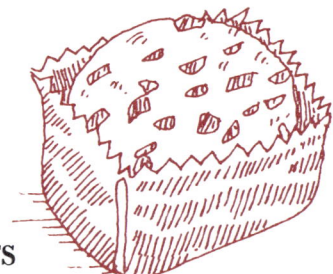

APLETS AND COTLETS

Tangy apricots and tart-sweet mountain-grown apples comprise this savory sensation . . . never-to-be-forgotten taste treats for the fruit-nut fancier. 1 pound 11 ounces in all! Half these exquisite candies are apple, half apricot—all delicately rolled in fresh, chopped nuts. Placed in bowls around living room and dining room, they'll be a sweet temptation to every passer-by—truly a tasty and gracious addition to all your entertaining. A gift so unusual, it's sure to be welcome and enjoyed by every special friend. $8.45 delivered.

MISSION PAK
Santa Clara, Calif. 95050

BUTTERY BLISS

Traditional favorite of continental bakers, these rich butter cookies melt on your tongue with a flavor so delicate and delicious you'll remember it for the rest of your life! Made from a centuries-old European recipe, they come frosted with delicately flavored pastel icings—some topped with yummy chocolate drops! Beautiful reusable tin will brighten any kitchen for many years to come!

    1-pound tin .... $5.95 delivered
    2-pound tin .... $8.95 delivered

MISSION PAK
Santa Clara, Calif. 95050

TOM 'N' CHERRIES

Rich, red cherries loaded with juice and covered with a thick, rich chocolate, make melt-in-your-mouth morsels that nobody can resist! Can't you picture the fun children will have when they see 12 ounces of these delicious drops "pouring" into their hands? These are California's tastiest cherries, hand-picked at their ripest and rushed to the chocolate maker for dipping. Every bite is a flavor delight! Wrapped in shiny red and gold, the decorative champagne bottle adds a sparkle to every party. A truly unique dessert that will keep guests talking and eating right through the last "drop." Available the year round. $5.95 delivered.

LEE ANDERSON'S COVALDA DATE COMPANY
P.O. Box 908
Coachella, Calif. 92236

DATES, DATE CONFECTIONS

If you have a sweet tooth, but are tired of consuming all the chemicals that go into candy, try dates, the natural, nutritious treat. Covalda brand dates are organically grown in the deep rich soil of the Coachella Valley in southeastern California. And the Anderson family's thirty years of experience stand behind the quality of the nine types of natural dates they sell. Also available: delightful date confections such as stuffed dates, date sugar, creamed dates, Date Chip TV Snacks, Date-Almond Confection, and many others.

Remember! Dates are one of the oldest treats known to man. And it has long been considered that a gift of dates is symbolic of a wish for a long life, both fruitful and happy. Try Covalda's 2½-pound Special Introductory Assortment for only $5.00 delivered.

Write the above address for a complete catalog, prices, and ordering information.

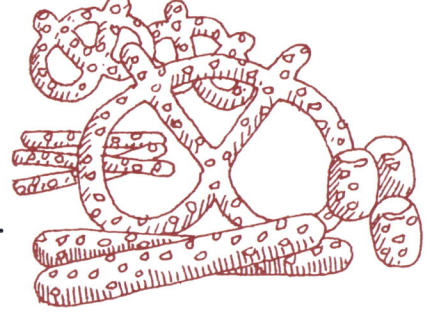

BACHMAN FOODS, INC.
P.O. Box 898
Reading, Pa. 19603

PRETZELS FROM THE PENNSYLVANIA DUTCH COUNTRY

Pretzels are always a favorite, for parties and between-meal snacks. Bachman provides you with a tasty assortment of pretzels—cheese stix, plain stix, bite-size bits, Dutch-style, thin-twisted, and petit pretzels. And the pretzel assortment comes in its own canister with an authentic Pennsylvania Dutch designed cover that can be used as a tray or a wall plaque. It weighs 5 pounds and costs $8.60.

For a complete description of Bachman's entire pretzel line, write to them at the above address.

BACHMAN FOODS, INC.
P.O. Box 898
Reading, Pa. 19603

TASTY MORSELS OF CHOCOLATE-COVERED PRETZELS

Mouth-watering goodness in bite-size pieces. It's a taste treat you won't forget. These mini-pretzels are drenched in pure milk chocolate, then packed in a colorful (and reusable) embossed tin. Each tin holds 1 pound of these delightfully unusual pretzels. $8.85 includes postage.

CONFECTIONS

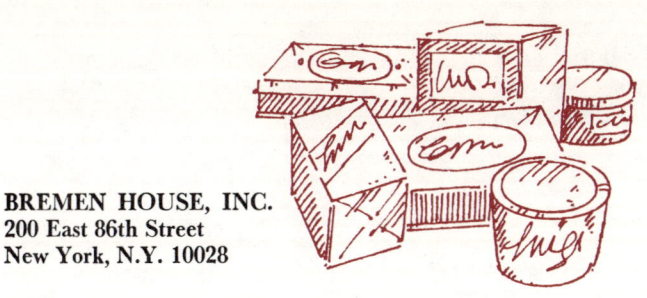

BREMEN HOUSE, INC.
200 East 86th Street
New York, N.Y. 10028

TRADITIONAL HOLIDAY SWEETS

Once Thanksgiving rolls around, most people's thoughts turn to the holiday season ahead. It means fantastic dinners, gourmet confections, and lots of love. What could enhance a holiday dinner more than a beautiful and delicious selection of traditional German cookies, cakes, Stollen, and marzipan? You can choose from Pfeffernuesse (spiced drops), 7 ounces for $1.10; Spitzkuchen (honey cakes with chocolate), 6 ounces for $1.75; Almond Speculaas, 7 ounces for $1.75; Zimtsterne (glazed cinnamon stars with ground hazelnuts), 3½ ounces for $2.00; Anissa (anise-flavored egg biscuit), 4½ ounces for $1.50; Corra (Lebkuchen hearts, chocolate filled with jam), 7 ounces for $1.50. All six packages for $9.00. A 26¼-ounce Christ-Stollen for $5.50 or buy a Marzipan Good Luck Pig in Cello Gift Box, 7 ounces for $3.50. Minimum order: $10.00.

See Condiments, Spices, and Syrups chapter for company description.

CARDINAL CANDIES
110 Lyman Street
Holyoke, Mass. 01040

POLKADOT PENNY CANDY BASKET, FOUR 'N ONE PINWHEEL

You are looking for a unique treat for those Halloween "goblins" who come knocking at your door. You need something new to decorate your child's birthday party table. Your organization gives a children's party every year. Cardinal Candies can provide you with the perfect treats.

Polkadot Penny Candy Baskets, available in blue or pink and white checked design and filled with an assortment of scrumptious candies, can be set at each table place or sent home with each child.

Four 'n One Pinwheels feature a hollow stem filled with colorful miniature sucking candies, and four colorful pinwheels. They can be used as table decorations, game prizes, or take-home gifts that will help the children remember the good time they had. Baskets are $4.20 per dozen in packages of 4 dozen. Pinwheels come in packages of 2 dozen at $4.80 per dozen.

EPICURES' CLUB
Elizabeth, N.J. 07207

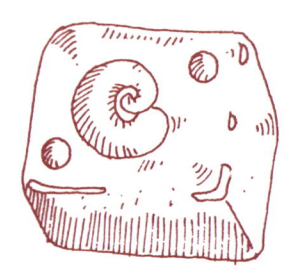

PARTY FOURS

To many Americans, petits fours are fancy cookies. To a Frenchman, petits fours are tea cakes variously frosted and decorated. There's a big difference, as any Frenchman will tell you. These Party Fours are like the French original. They look like candy, but inside is cake . . . and cake such as you wouldn't believe it possible to produce except in your own home—and you'd have to be an expert baker to even come close. The big problem in developing these Party Fours was to improve the keeping quality without sacrificing the taste of the product in any way. It took three years of painstaking work to come up with the right formula. And those three years were spent by an expert—a European who has been making candies and cake for forty years. Party Fours will stay fresh for six weeks under ordinary conditions and for six months in your refrigerator or freezer. 20 ounces (40 pieces)—$4.95 delivered. 40 ounces (80 pieces)—$9.25 delivered. 10 ounces (20 pieces)—$3.65 delivered.

See Meat, Fish and Poultry chapter for company description.

AMISHMAN
Mount Holly Springs, Pa. 17065

APPLE BUTTER

Do you develop a craving for apples when the fall months roll around? Then you should try this delicious, homemade apple butter. Made of a combination of apples and fresh-fruit apple cider, it's skillfully blended in a wooden barrel with a copper lid and coil. You'll love it on pancakes, toast, muffins, rolls, biscuits, waffles—in fact on almost anything. And this sweet full-flavored apple butter is made by a Pennsylvania Dutch family from a traditional recipe handed down since 1734. It's today's way of getting yesterday's old-fashioned goodness. 18-ounce glass jar—$2.00.

CONFECTIONS

SMUCKER'S
Box 55
Orrville, Ohio 44667

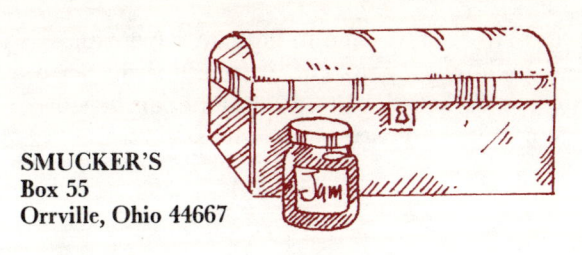

SMUCKER'S GIFT BOXES

If you love rich, flavorful jams and jellies, here's the perfect gift for you: a wide assortment of Smucker's gift boxes. These distinctive, colorful arrangements come packed with a variety of Smucker's own jams, jellies, and preserves. The Stars & Stripes Gift Chest, for instance, is packed with twenty 6-ounce jars, including peach butter, mint-apple jelly, strawberry preserves, chocolate fudge topping, candied relish (8 ounces), grape jelly, and many others. And when the jam's gone, the chest remains—to carry a thousand different things. Each Stars & Stripes Gift Chest weighs 23 pounds and costs $23.95.

ASTOR CHOCOLATE CORPORATION
48–25 Metropolitan Avenue
Brooklyn, N.Y. 11237

CHOCOLATE LIQUEUR CUPS

Save the best for last at your next dinner party. Your guests will certainly remember you for it. These are delicious bittersweet chocolate cups "glasses" that you fill with your favorite liqueur. When your guests have consumed the liqueur, they eat the chocolate "glass" cup! What fun. Your guests will "sip and nibble" their way to the end of a long-remembered meal. Cups come with gold foil holders, and are gift-packaged 12 to a box. Can be stored indefinitely in a cool place. Minimum order: 3 boxes—$6.50.

ASTOR CHOCOLATE CORPORATION
48–25 Metropolitan Avenue
Brooklyn, N.Y. 11237

PETIT CHOCOLATE DESSERT SHELLS

The beauty of these delightful little bittersweet chocolate dessert shells can only be enhanced by your own personal decorating touches. Fill them with diced gelatine, flavored whipped cream, grapes, fruit marinated in liqueur, ice cream, etc.—the possibilities are endless. And because they're petit, it makes them perfect for teas and light desserts. Shells come packaged 10 to a box and can be stored indefinitely in a cool place. Minimum order: 3 boxes—$7.25.

EDWARDS PRODUCTS COMPANY
P.O. Box 411
Zebulon, Ga. 30295

FANCY PECANS

If pecans are your weakness, you'll love this taste treat from the heart of pecan country. Edwards fancy pecans have a delicate flavor that will tickle your tongue and tease your taste buds. They're a delicious way to spark up your pies, cakes, cookies, and salads. Selected from the finest available pecan meats, they're specially chosen for size, texture, and flavor. Use them to make a refreshing ice cream sundae, a luscious pecan pie, or a nutritious, wholesome afternoon snack. Edwards fancy pecans—for the discriminating palate. Price per pound is only $3.60, including shipping.

WISCONSIN CHEESEMAN
P.O. Box 1
Madison, Wis. 53701

HAPPY EATING

Do you rack your brain every time you have to give a gift? Are you bored by giving the usual ties, handkerchiefs, perfume, costume jewelry, liquor as personal and business gifts? The Wisconsin Cheeseman may be able to solve your problem. An unusual assortment of foods has been selected and put together in the "Happy Eating" package. It offers fourteen different items: pecan log, coconut fudge log, all-beef sausage, summer sausage, cheese logs rolled in parsley and in paprika, aged Cheddar and American cheese bricks, mint and chocolate tortes, richly decorated butter cookies, and strawberry candy. All for only $14.95, delivered with a personal gift card on the date you specify.

CONFECTIONS

AUNT SALLY'S SHOPS
810 Decatur
French Market
New Orleans, La. 70116

WHOLE PRESERVED LOUISIANA STRAWBERRIES

Freshly picked, these firm and luscious Louisiana strawberries are preserved whole and packed in 3-pound old-fashioned wooden pails, and mailed directly to you or your friends. They can be used to fill and decorate cakes and cupcakes for special occasions, to top ice-cream desserts, to accompany elegant brunches. A perfect gift for those of you who prefer high-quality preserves. 3 pounds, packed in wooden pail—$9.95.

AUNT SALLY'S SHOPS
810 Decatur
French Market
New Orleans, La. 70116

ORIGINAL CREOLE PECAN PRALINES

A combination of the French confectionary art and gracious ante bellum New Orleans living are brought to mind when you bite into an Original Creole Pecan Praline. An ideal after-dinner sweet. An added touch of elegance to formal luncheons, dinners, and club receptions. A unique gift for gourmet friends with a sweet tooth. You can order a box of one dozen, packed in charming miniature cotton bales, for $4.75. Gift card included.

CALIFORNIA ALMOND GROWERS EXCHANGE
1802 C Street
P.O. Box 1768
Sacramento, Calif. 95808

EL CAPITAN ALMONDS

If you love crisp, crunchy almonds, then you'll love this special assortment of *six* varieties of almonds. Attractively and colorfully packaged, this delicious assortment is the perfect way to satisfy your almond craving. Its six different kinds of fresh-roasted almonds are hickory-smoked, roasted salted, blanched salted, cheese, onion-garlic, and barbecue. You'll love them for parties, school lunches, picnics, evening snacks, or anytime people gather together. Known as the El Capitan, it makes a delightful gift, too. It's a treat that's sure to be remembered. Cost for six 6-ounce tins in only $6.50 delivered.

PACKING SHED
P.O. Box 11
Weyers Cave, Va. 24486

HOME-COOKED SALTED PEANUTS

If you like peanuts—without the preservatives—you'll enjoy the delicious taste of these specially selected Virginia Jumbos. Free from the additives and preservatives found in most commercial brands of peanuts, they are prepared in small batches for peak flavor. Home-cooked salted peanuts are specially prepared through a process of shelling, water blanching, cooking in oil, and salting—and are then packed in tins for immediate shipment. Because they're specially prepared, home-cooked salted peanuts should be kept in your refrigerator—or in the freezer for longer periods of storage. A 2-pound 14-ounce tin costs $5.95, plus $2.00 for shipping. A 1-pound 4-ounce tin costs just $3.50, plus $1.30 for shipping.

WILLIAMS-SONOMA
Mail Order Department
532 Sutter Street
San Francisco, Calif. 94102

A DIFFERENT KIND OF JAM

Frankly, this jam is addictive! No pound for pound of fruit and sugar here, this is fruit simmered with yet more concentrated fruit and a minimum of sugar and it is absolutely marvelous. Made lovingly in Ireland and packed in attractive old-fashioned reusable stoneware crocks. Black currant, strawberry, and raspberry jam, 16-ounce crocks, the assortment of three is $9.00 plus $1.75 for shipping.

See Kitchen Utensils chapter for company description.

CONFECTIONS

FIGI'S, INC.
Marshfield, Wis. 54449

DIETETIC JUBILEE

A large assortment of the finest products available for those who wish to avoid sugar. The colorful gift box contains a variety show of sugarless delights featuring 6 ounces of colorful cookies; a 4½-ounce box of chocolate-covered fruit and nut mix; 8 ounces of sugarless hard candy; 4 ounces Cheddar cheese wheat germ wafers; and two 7/10-ounce chocolate wafer bars. $7.50.

See Cheese chapter for company description.

FIGI'S, INC.
Marshfield, Wis. 54449

GINGERBREAD HOUSE YOU CAN EAT

Children's faces adorn the window and St. Nick opens the door of this cozy cottage. Constructed of traditional holiday gingerbread, decorated with tasty icing and colorful candies. Enchanting 3-pound 5-ounce house arrives fully assembled and stands 8″ high on a 10¾″×7½″ base. $9.50.

FIGI'S, INC.
Marshfield, Wis. 54449

A BOXFUL OF CHEESE AND OTHER GOODIES

Everyone likes variety, especially when it's tastefully packed in an exciting and appetizing gift assortment. An impressive selection of the most popular varieties to suit every taste. The cheeses are 4 ounces of caraway, bleu, and smoked Edam; two 4-ounce Edam, Colby, Cheddar, and Gouda; 5¼-ounce cuts of salami cheese and Colby; six 1-ounce links; two 1-ounce Kummel; two 5-ounce glacéed fruit; 2¾ ounces butter cream and mint cream tortes; 6 ounces butter cookies; 5 ounces petits fours; a 5¾-ounce pecan log; and 4 ounces all-beef summer sausage sticks. Truly an impressive gift assortment for your many friends. $18.95.

See Cheese chapter for company description.

HARRY AND DAVID
Medford, Oreg. 97501

CONFECTIONS 'N' FRUITS

Here's a mouth-watering treat that you'll find difficult to pass up. From the orchards of Oregon come Harry and David's world-famous apples and pears. But that's not all. They've also included a tempting selection of nuts, dates, and other candied fruits. Over 7½ pounds in total. $12.95 delivered.

See company description in this chapter.

HARRY AND DAVID
Medford, Oreg. 97501

A MINI-SPRUCE DECORATED WITH CONFECTIONS

Oregon north woods beauty and fresh evergreen scent in a perfectly formed, live miniature Christmas tree. This bushy, symmetrical dwarf Alberta spruce, already four years old and over a foot tall, is prepotted and completely decorated. And wait until your friends see it all aglitter and aglow with twenty-two little individual gifts: tiny peppermint candy canes, milk chocolate foil-wrapped ornaments,

CONFECTIONS

and two kinds of homemade Christmas cookies—spicy iced gingerbread and rich sugar cookies. This hardy little tree can be planted outside in the spring. Compact, trouble-free—it always retains its perfect dwarf pyramid shape, growing slowly to about seven feet tall. $16.95 delivered.

HARRY AND DAVID
Medford, Oreg. 97501

HARRY'S SUGARPLUMS

More than a chocolate—yummier than plum pudding—better than candy! Heavenly nougats of crisp English walnuts, zingy apricots, and plump sweet Medjool dates, the richest and costliest dates of all, finely chopped and combined just so, then dipped—half of them into creamy milk chocolate, and half into extra-rich dark chocolate. Over a pound in a fancy gift box. You won't meet up with confections like these in stores. More healthful than candy too, because they're chock full of the natural sweetness of fruit. There's never been such a heavenly concoction—and at such a nonextravagant price. $8.45 delivered.

HARRY AND DAVID
Medford, Oreg. 97501

PETITS FOURS

The dessert served by the French Foreign Legion . . . chosen cutest sweet of the week by Cordon Bleu . . . considered most acceptable in social circles. All that, and they're absolutely exquisite eating, too! Petits fours look like fancy chocolates, all iced and hand-decorated. But inside they're light and fluffy individual layer cakes. Made from the finest quality ingredients—no artificial preservatives. Stay fresh in refrigerator or freezer for up to six months. Get several—have 'em on hand for elegant emergencies. Net weight 1 pound 1 ounce—$7.45 delivered.

ABOUT THIS COMPANY

Since 1936, Harry and David have promised to always please you—no matter what. Their town lawyer, who used to be a good pear-picker, says farmers can't guarantee anything because of hailstorms, wind, drought, crop failures, freezes, mistakes, hazards of farming, birds and bees and everything else. You'll find his fine print on their order blank, but remember—they guarantee complete satisfaction in every respect or your money back.

PLUMBRIDGE
33 East 61st Street
New York, N.Y. 10021

CHOCOLATE TRUFFLES

If you enjoy chocolate in all its varieties, you'll have a mouth-watering good time with Plumbridge chocolate truffles. Skillfully prepared using the finest recipes available, these delectable candies are filled with an old-fashioned goodness that will keep you coming back for more. And they come in a variety of flavors and coverings—including cream mints, caramels, apricots, coconuts, and marzipans—all covered with thick, rich dark chocolate. Price is just $7.00 per pound.

GRACE A. RUSH
3715 Madison Road
Cincinnati, Ohio 45209

GINGER JOHNNY CAKE

Ginger has been prized by man for centuries. It is one of the fabled spices of the Orient, the object of many of the earliest voyages of exploration. It was among the treasures brought to Venice in 1295 by the famed Italian traveler Marco Polo, after a twenty-four-year stay in the Orient. It is no less highly prized today . . . it is used as a basic ingredient in almost all Oriental cooking. It's a fabulous experience to taste this marvelous cake packed in a beautiful 1-pound gift tin. There's nothing like it in the world—it's brand-new. The ginger buffs will love it because it contains tiny pieces of the most succulent Oriental ginger—see and taste for yourself. Then you will want to send this enchanting gift to the friends you love best. 1 pound—$4.50 postpaid U.S.A.

WILLIAMS-SONOMA
Mail Order Department
532 Sutter Street
San Francisco, Calif. 94102

CHRISTMAS STOLLEN

Stollen, a delectable holiday cake that originated in Dresden, is traditionally exchanged as a Christmas gift all over Germany. This cake, which gets even better as it ages, consists of just enough cake to hold together a fine rum-flavored assortment of raisins, crystallized fruit, and almond paste, and is liberally coated with confectioners' sugar. 1-pound 10¼-ounce cake, boxed—$6.50.

See Kitchen Utensils chapter for company description.

VERMONT COUNTRY STORE
Weston, Vt. 05161

A TRULY SUCCESSFUL GOURMET'S FRUITCAKE MADE FROM AN OLD RECIPE. UNSURPASSED IN RICH, OLD-FASHIONED GOODNESS

If you are stumped for a last-minute gift, try this delicious fruitcake, specially packed in a plain, round tin box, containing a full 3 pounds of cake. And what cake! Walnuts and candied cherries from France, almonds from Spain, pineapple from the Caribbean, whiskey from Kentucky, and plenty of rum, brandy, wine, raisins, and dairy products—all creating a rich cake loaded with extra goodies and enhanced by the good flavor of bonded Bourbon. No one could resist such a treat. Cake like this is acceptable anytime. You can't go wrong buying several of these wonderful cakes for your family and friends. $10.95.

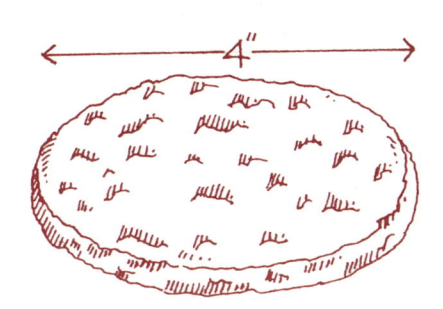

JUMBO WHOLE-MEAL COOKIES

These big cookies will measure about 4″ in diameter. Because the stone-ground meal is rich in protein and vitamins, one of these big ones is great for the kids as a snack or dessert because it's good for them. And when you serve them at tea or a party, they are pretty sure to meet with enthusiasm. Now please understand that these cookies (being so big) may become a bit brittle. Depending on weather and how they travel to you, some may be cracked or broken. But if you are willing to take that chance, you can buy 15 assorted cookies. $4.00.

FRESHLY BAKED WHOLE-GRAIN BREAD

The Vermont Country Store made this bread famous all around the area. They have customers who drive fifty miles to buy it, because it's made from their famous stone-ground 100 per cent whole-grain wheat meal. And, as everyone knows who has used this stone-ground whole-grain wheat meal, which they have sold by mail for many years, it contains all of the hard northern spring wheat, rich in protein, trace elements, and vitamins, nothing left out, nothing put in. People relish this bread for the homemade taste. Others, who are watching weight, eat it because one loaf furnishes several times more nourishment than lifeless white bread, which is often so fluffy and empty. And it has all the natural goodness, with no synthetic additives. They pack three 22-ounce loaves in a carton and ship them to you, fresh from their ovens. (They don't recommend ordering it if you live west of Texas.) $3.15.

CRYSTALLIZED CANDIED GINGER AND CANDIED ORANGE PEEL

These two old-fashioned confections have been enjoyed by generations of Americans in cooking and as a sweet treat. The ginger, in large pieces, is coated with sugar syrup, the real strong kind our grandmothers remember. The orange is thin strips of the peel with the same coating. The bittersweet of the peel, combined with the sugar, gives a strong orange essence in cooking and interesting taste for eating or serving with tea. Each packed 1 pound to a box.

    Crystallized Ginger. . . . . $3.25 a pound
    Crystallized Orange . . . . $2.75 a pound

CONFECTIONS

OLD-FASHIONED APPLE BUTTER

This apple butter is the traditional thick, spicy, old-time favorite so delicious on toast or muffins and pancakes, too. It's tasty with pork, chicken, or turkey. Good because it's made from choice, fresh apples, cider, brown sugar, cinnamon, and cloves. It's different because first it's cooked in a pressure cooker, then in the old-fashioned open kettle until it comes out a dark, thick, rich condiment. The most natural country product you'll be likely to find; no artificial ingredients. Large 28-ounce jar—$1.25.

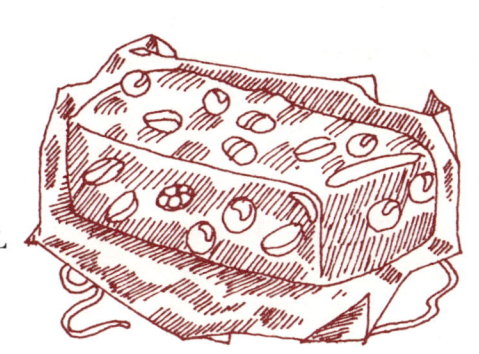

**MERV GRIFFIN'S CARMEL
CRAFTSMEN'S CATALOG**
P.O. Box 1600
Carmel, Calif. 93921

BRANDIED DATE NUT CAKES AND FRUITCAKES

South of Carmel by about twenty miles, in the rugged Big Sur region of California's incomparable coast line, is located the New Camaldoli Hermitage. As is true with most monasteries, the monks at the Hermitage create different items to support their way of life. One of the most incredible things made here, as we have found, is their brandied date nut cakes and fruitcakes. After perfecting this age-old recipe and process, the Hermitage monks have created what can only be described as a treat to the palate of even the most discriminating gourmet. The monks do everything right there on their vast property from chopping the nuts to wrapping the final cakes for sale. The balance between brandy, nuts, and fruit achieves an exquisite delight that is sure to spark the taste buds of each and every lucky sampler. Even if you're not a fruitcake lover, you'll find yourself craving these two delightful cakes time and time again.

BOXED: Brandied Date Nut Cake, 3 pounds $10.95
Fruitcake, 3 pounds $10.95

BEVERAGES: COFFEES AND TEAS

Probably no other food has become a way of life in the same manner of coffee and tea. In America, the first cup in the morning has become a near necessity, while in England, the entire day is scheduled around the afternoon tea.

The origin of coffee is rich in legend. One story tells of an Arab holy man who lived in the ninth century. He frequently fell asleep in the middle of his prayers, and he blamed his drowsiness on the lessening of his convictions. One day, he met a wise shepherd who recommended a "magic" berry that, when eaten by his goats, kept them continuously awake and in a festive mood. Anxious to complete his prayers, the holy man partook of this strange berry. And he also found that he could stay awake and be in a festive mood.

Interestingly, coffee was first prepared as a food, not a beverage. Coffee didn't become a hot drink until A.D. 1000 when the Arabs learned to boil it. It eventually affected every aspect of life in the Near East. So important was coffee that Turkish wives could legally divorce husbands who didn't supply them with this beverage! However, the people of Venice were responsible for finally spreading coffee around the world.

Most coffees are grown in South America, Central America, Jamaica, Java, and Africa. Beans can be roasted many different ways and blended to produce an endless variety of taste, from expresso to mocha to good old black.

Tea is the most universally consumed beverage in the world and has progressed far beyond the orange pekoe variety. Ceylon teas, Chinese oolong tea, and even ginseng tea (which has a mystic all its own) are now available through the mail. It would be a pity to deprive yourself of so much variety.

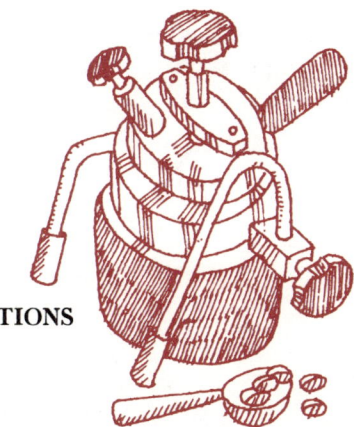

FERRARA FOODS AND CONFECTIONS
195–201 Grand Street
New York, N.Y. 10013

CAPPUCCINO

If coffee's your thing, you'll love Ferrara cappuccino. Made from a special blend of carefully selected, high-grade coffees, this delicious Gourmet Brown Roast is perfect for the connoisseur. Serve with chocolate mousse, biscuits and honey, vanilla ice cream, wafer cookies, or just by itself. No matter when you drink it, its warm, refreshing taste will delight you—and your guests—every time. Cappuccino is the finishing touch for an afternoon get-together, elegant dinner party, or evening tête-à-tête. A 1-pound can costs just $2.00.

ABOUT THIS COMPANY

Ferrara Foods and Confections is a national business. In addition to the large retail trade which continues to grow, Ferrara packages and ships its own brands of coffee, both espresso and cappuccino, petit rhum babas, candies, and hundreds of other Italian delicacies to department and specialty stores from coast to coast. Send for their flier.

GRACE TEA COMPANY, LTD.
799 Broadway
New York, N.Y. 10003

SUPERB AND RARE TEAS FOR THE CONNOISSEUR

A true connoisseur of tea enjoys a wide variety of types and blends, just as the true lover of wine. Grace offers you the finest teas available, chosen for taste and quality.

Formosa Oolong Supreme is the champagne of teas, the personal choice of professional tea tasters. This great tea of delicate flavor and elegant bouquet is handpicked only once a year when the flavor is at its peak. $4.00 the half pound.

Winey Keemun English Breakfast is a fine silvery leaf, extra-slow hand-fired to create a tea of rarity and superb quality. Its smooth winey flavor makes it a must for tea lovers. $4.00 the half pound.

A unique treat for yourself, the perfect gift for your tea-loving friends.

Write the above address for order forms and a complete catalog of Grace rare teas.

NICHOLS GARDEN NURSERY
1190 North Pacific Highway
Albany, Oreg. 97321

MORMON OR SQUAW TEA

It's always been claimed that tea has medicinal properties. Here's one that's extra special. Indians have used this tea for centuries, crediting it with almost miraculous medicinal and tonic properties. The tea is a stimulant and reputed to be

BEVERAGES: COFFEES AND TEAS

good for kidney ailments and stomach disorders, as a blood purifier, and an excellent general tonic. The early Mormon pioneers used the tea extensively for a variety of ailments, and claimed that it helped save many lives during one of their cholera epidemics. Many people today, however, claim the tea has a therapeutic value for the relief of arthritis. The tea foliage is taken from a small shrub that resembles a miniature pine tree found growing in the high deserts of Nevada, Utah, and California. The pine-needle-like foliage is gathered at elevations of five to eight thousand feet, where nature is pure and clean, untainted by chemical pollutants. This tea has the most pleasing flavor of all medicinal teas—never bitter—and brews into a beautiful pink color. It is delicious with a little honey and the twist of a small lemon wedge. 4 ounces—$1.75. 12 ounces—$4.25. 4 pounds—$18.00.

See Cheese chapter for company description.

EMPIRE COFFEE AND TEA CO.
486 Ninth Avenue
New York, N.Y. 10018

HAITIAN COFFEE BLEND II

Haiti's warm climate and fertile soil produce a deep rich coffee. Often, it's "too strong" for American taste. However, if you're a real coffee lover, you'll be impressed. Haitian coffee tends to be quite bitter in the less expensive blends. The Empire selection is made only with Grade 1 robust beans grown on the south (sunny) side of the island. A pound fresh-ground is $3.98 in an attractive burlap sack.

Empire imports and packs over 168 types of coffees (and teas), including the rare Formosa green, Tahitian yellow, and Columbia gold. Catalog is free.

JAMES G. GILL COMPANY, INC.
204–210 West 22nd Street
Norfolk, Va. 23517

CHINESE COFFEE

If you've always thought of the Chinese being famous for their tea—guess again. A new variety of coffee has joined Gill's famous gourmet list. It's Chinese coffee.

This superb coffee is derived from a species of coffee bean known as China Arabica, which is grown in the southern Chinese province of Yunnan near Burma. After roasting, it has a deep, rich-brown color and a full-bodied, low-acid taste that coffee lovers will find pleasing. But as Gill's vice-president describes it, this coffee has something more: "a mysterious kind of flavor that typifies the Orient somehow." Why not experience some of China's select coffee today? Cost is only $1.89 for a 1-pound sack.

ABOUT THIS COMPANY

The James G. Gill Company, Inc., of Norfolk, Virginia, has served the specialty food trade as an importer and roaster of fine coffees since 1902, when Gill installed the first roaster in his plant near the docks of that port city. Today, four generations later, the family is the leading processor of gourmet coffees on the East Coast, with over 30,000 pounds being produced daily, and offers the nation's most complete line of coffees and teas for the connoisseur. Under their First Colony label, Gill's markets a superior collection of almost thirty unblended varieties of coffee as well as numerous blends. Among these rare offerings are the first major import of Jamaican Blue Mountain in decades, one of the very few genuine Arabian Mochas, and three varieties of decaffeinated bean. Rare China Yunnan Arabica coffee is the latest addition to the line. You can also choose Brazilaro, Mocha Java, Chicory, Vintage Colombian, Venezuelan Tachira, Ethiopian Harrar, Plantation Kenya, Guatemala Maragogipe, Costa Rican Tres Rios, and many others. This unparalleled selection of coffees is complemented by fine bulk teas in over thirty varieties, purchased by Gill's in all the major tea markets and auctions throughout the world. One of the few authentic Golden Tip Darjeelings (TFGOPI) available in this country is marketed under the First Colony label. For complete descriptions and price list, write the above address.

CAPRICORN COFFEES
353 Tenth Street
San Francisco, Calif. 94103

WHOLE BEAN COFFEES, EXOTIC TEAS

Now you can order directly the same fine-quality coffees and teas formerly available only to restaurants and specialty shops. Among these specialty brews is Ethiopian Mocha Harrar—a super-succulent blend of rich, dark coffee beans picked from the lush trees of this African paradise. It's a deliciously warm beverage you'll love anytime. And Ethiopian Mocha Harrar is shipped whole bean

BEVERAGES: COFFEES AND TEAS 141

(unless grind is specified) to retain its freshness until you are ready to grind and brew the best pot of coffee you've ever tasted. Cost is $3.10 per pound.

You can also choose from South and Central American coffees, Caribbean coffees, Mid and Far East coffees, as well as sixteen exotic teas and four old-fashioned herb infusions. For further information and prices, write the above address.

ABOUT THIS COMPANY
Capricorn Coffees imports and *daily* roasts their coffee. They have been serving specialty coffee stores and restaurants throughout northern California and the West since 1963.

NORTHWESTERN COFFEE MILLS
217 North Broadway
Milwaukee, Wis. 53202

BLACK TEA
Most commercial teas and many specialty teas, including some of Northwestern's selections, are blends. A good blender seeks to balance the taste of the brew, but sometimes must sacrifice the taste of a desirable individual tea in order to make the blend work together as a whole. Northwestern's unblended teas offer you the opportunity of creating blends that may best meet your special taste requirements. Their blended teas include some brisk flavored teas, some rich, thick teas, and some teas of special flavor characteristics. Black tea is a fully fermented tea. First, the leaves are rolled, which accelerates a natural fermentation process by releasing oxidizing enzymes. The fermentation is stopped by heating the leaves at high temperature. The best tea is heated, or fired as it is termed in the tea trade, at precisely the right moment, with the leaves being neither under- nor overfermented. The type of firing process is sometimes used to describe the tea, such as basket-fired or pan-fired tea.

Assam Their strongest tea, pungent and brisk in flavor. Grown in the Assam district of eastern India, this tea has a mellow body that makes it outstanding, either in its own right or when its distinctive character is added to any blend. ½ pound—$2.40. 1 pound—$3.80.

Backsettler Blend Like the teas for which the mountain men traded their furs, this tea is hearty and brisk with the combination of flavor, body, and aroma that have made black teas the world's favorite. It fits nearly all tea-drinking occasions well and is forgiving to brewing mistakes. ½ pound—$2.40. 1 pound—$3.80.

Darjeeling, Second Flush Grown in the foothills of the Indian Himalayas, Darjeeling has a reddish color and a very fragrant aroma. Its exciting bouquet is unique among teas. The finest grade of Darjeeling is the rarest and costliest of the black teas, the second flush or picking of the best tea bushes. ½ pound—$2.80. 1 pound—$4.80.

See company description in this chapter.

NORTHWESTERN COFFEE MILLS
217 North Broadway
Milwaukee, Wis. 53202

JASMINE BLOSSOM BLEND

If you enjoy exotic teas, here's one to add to your collection: Jasmine Blossom Blend. This unique Oriental tea, from Mainland China, is a blend of scented oolong and delicate whole jasmine blossoms. It's a truly distinct beverage that's perfect for elegant luncheons, gourmet dinners, afternoon teas, or even as a "nightcap." And because this highly flavorful tea has only recently become available, it's an exciting new way to please your guests. ¼ pound—$1.50. ½ pound—$2.50. 1 pound—$4.00.

ABOUT THIS COMPANY

The Northwestern Coffee Mills' tradition began in 1875 as Milwaukee's first coffee and tea merchant. Emphasis has always been on keeping the business personal and quality-oriented. They have been at their present location roasting coffee since 1918. All of their products, both those manufactured by them and those distributed for others, are held to the same high standards.

VERMONT COUNTRY STORE
Weston, Vt. 05161

TEAS FROM CHINA

English Breakfast is actually a fine black China tea traditionally preferred by the English for breakfast. Medium black in strength and color. ½-pound tin—$3.25.

Formosa Oolong is a superior-quality, aromatic, fine-flavored tea, often called Champagne of Teas. For folks who have never tried any vintage tea, we recommend they start with this one. ½-pound tin—$2.75.

Jasmine is a delicate light tea with a hypnotic aroma due to lovely jasmine petals that make it a special treat for festive occasions. ½-pound tin—$3.50.

Pinhead Gunpowder is the extremely rare green China tea, long a Mandarin favorite for the tea ceremony. Almost unknown here. Its leaves are tiny, tightly rolled, and tender, with a pleasing aroma. ½-pound tin—$3.25.

China Pouchong is the extra-choice green tea of extremely rare vintage and seldom brought over. Raised for centuries in China, it is now grown in some parts of southern Vietnam at the highest elevations. The tiny succulent leaves, carefully selected, are silver-tipped and hand-rolled and exude a delicate and captivating aroma and taste. ½ pound—$2.95.

Russian Caravan was used for centuries by the hardy caravans that trekked across China into remote parts of Asiatic Russia. A heavy-bodied black tea with authority. ½-pound tin—$2.75.

Lapsang Souchong is the famed smoky black tea in aroma and flavor. This is such a rare full-bodied tea, usually reserved for special occasions. The pleasing smoky aroma and taste are remarkable. ½ pound—$3.25.

DECAFFEINATED COFFEE IN FRESHLY ROASTED BEANS

Some folks believe that coffee drunk at night keeps them awake. For such people, there is something new—a decaffeinated coffee *in the bean* (not instant) that can be ground fresh and made fresh. The Vermont Country Store roasts the green beans right on the premises and they will also grind it for you. Or better still, grind it in your own home coffee mill. Decaffeinated coffee, 1 pound—$3.25. (1 pound minimum order.)

WELL-PRICED, UNBLENDED COFFEES

In 1976 bad weather in Brazil destroyed much of the crop which reduced the world-wide supply of coffee. As a result, the price of Brazilian Santos went way up and immediately all the other coffee countries raised their prices. Also, it seemed as if all the specialty stores raised their prices way up as well. Coffee became very expensive. However, the Vermont Country Store still keeps their coffee prices in line. Here are some basic coffees which you can purchase to create your own blend or use as highly individual coffees.

Java is a rare aromatic coffee of great flavor, with a rich body. $2.95 a pound.

Santos The best quality Santos from Brazil is heavy-bodied and has rich winey flavor; usually used as a major base for most famous coffee blends. $2.95 a pound.

Colombia comes from that South American country known for its top-quality coffees. A distinct clear flavor and body, Colombia is nice and mellow. $3.50 a pound.

Guatemala Antigua is raised on the 8,000-foot-high mountain Fincas in Guatemala and is used to pep up coffee blends. It is considered one of the most expensive and is produced in small harvests. Has a good body with a tantalizing aftertaste like fine wine. $2.95 a pound.

NORTHWESTERN COFFEE MILLS
217 North Broadway
Milwaukee, Wis. 53202

COFFEE BLENDS

Northwestern Coffee Mills' blends are formulated to suit widely varied taste preferences. To create a blend, they mix two or more coffees to gain a different and desirable taste, without primary regard to the cost of the ingredients. Their goal is to find a combination of coffee aroma, body, and flavor that is more satisfying to many people than any of the individual coffees constituting the blend. These blends generally use substantial percentages of low-quality coffee such as the Robusta varieties, which have so little taste and aroma to recommend them that they are seldom sold unless hidden in blends of better coffees. They use only the very best Arabicas.

American Breakfast Blend   They have blended this morning coffee according to an old family formula. It has a full flavor, mild body, and delicate aroma for those who prefer to drink several cups at a sitting. 1 pound—$2.45. 5 pounds—$11.25.

Fancy Dinner Blend   Their own creation of a bold coffee for ending the evening meal on a high note. Its aroma is strong from a touch of dark roast and its body has a full consistency complementing this blend's hearty taste. 1 pound—$2.50. 5 pounds—$11.50.

Hawaiian Kona Blend   Nurtured by the volcanic soil on the slope of Mauna Loa, Kona coffee has a superior and distinctive mild flavor. They have balanced it with the aroma of rich Central American coffees. 1 pound—$2.80. 5 pounds—$13.00.

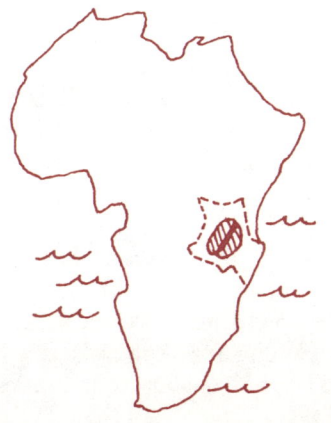

Kenya XX Blend   Grown in the East African high country. The deep aroma and lively taste of this blend is an unusual coffee experience. 1 pound—$2.75. 5 pounds—$12.75.

Mocha Java Blend Perhaps the most famous of coffee blends because of the near perfect match of Java's creamy richness and Mocha's piquant taste. 1 pound—$2.55. 5 pounds—$11.75.

Stapleton Restaurant Blend The favorite of Milwaukee's finest restaurants for nearly one hundred years. It is deeply satisfying to most taste preferences because of its full body, easy taste, and forgiving nature when kept heated for long periods of time. 1 pound—$2.40. 5 pounds—$11.00.

See company description in this chapter.

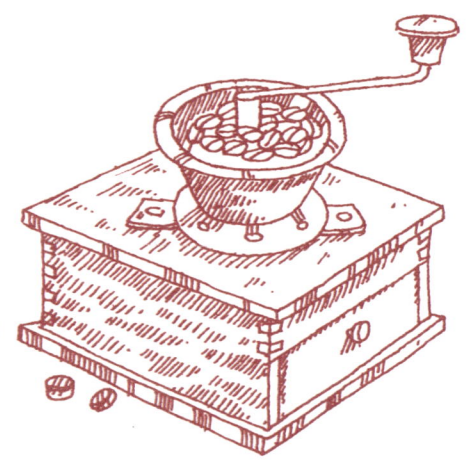

NORTHWESTERN COFFEE MILLS
217 North Broadway
Milwaukee, Wis. 53202

DARK ROAST COFFEES

An important variable in the taste of coffee is the length of time used in roasting the raw coffee beans. Most of Northwestern coffees are a rich brown American roast, neither light and tasteless nor burned and bitter. They create dark roasts by keeping their coffee in the oven longer than the regular American roasts, with the flames set lower to roast evenly without burning. The resulting rich, extra-roasted taste and glossy, dark beans distinguish these dark roast coffees.

French Roast The French, or Continental, Roast is a deep brown, with the coffee's oil brought to the surface of the bean. This roast is the favorite of many Central and South American coffee drinkers as well as European consumers. Although only slightly darker than American roast, French Roast has a heartier body and sharper taste. 1 pound—$2.45. 5 pounds—$11.25. ½-pound quantities available at half the 1-pound price.

Espresso Roast Also known as Italian Roast, the espresso roasted bean is a black-brown and glistens with coffee oil. As with the French Roast, espresso has great potential in recipes as well as by itself. Warm milk, hot chocolate, or a twist of lemon is recommended company for espresso's penetrating robust flavor. 1 pound—$2.55. 5 pounds—$11.75. ½-pound quantities available at half the 1-pound price.

See company description in this chapter.

BEVERAGES: COFFEES AND TEAS

MOTHER'S GENERAL STORE
P.O. Box 506
Flat Rock, N.C. 28731

HERBAL TEAS

Nothing rounds out the Good Life on the Complete Homestead like a steaming cup of herb tea (sweetened with honey) in front of a roaring fireplace on a cold winter's evening. Four of the most delicious, healthful, and popular herb teas are listed here. Ounce for ounce, bulk teas are much less expensive and go so much further. Order one of these herb teas and see for yourself.

Peppermint (6 ounces)	$1.50
Rose Hip (16 ounces)	$1.98
Camomile (6 ounces)	$1.98
Sassafras Bark (6 ounces)	$1.75

See Food Kits chapter for company description.

PAPRIKAS WEISS IMPORTER
1546 Second Avenue
New York, N.Y. 10028

INTERNATIONAL TEA FOR EVERY TASTE

Imported, selected, blended, and checked by **Paprikas Weiss**, this is an assortment of the finest teas in the world—each with its distinctive aroma. Experiment with the infinite variety of tea flavor.

Chinese Restaurant Tea   Delicate and fragrant. ¼ pound—$2.98.

Earl Gray's Tea   World-famous blend of rare fancy-scented teas. ¼ pound—$3.98.

English Breakfast Tea   Strong-flavored blend of Indian and Ceylon teas. ¼ pound—$2.98.

Russian Samovar Tea   Brisk, hearty taste meant to be brewed at full strength. ¼ pound—$2.98.

Irish Breakfast Tea   A pungent, dark amber brew of Assam and Ceylon teas. ¼ pound—$2.98.

Black Indian Tea   Indian and Ceylon teas combined for delicate flavors. ¼ pound—$2.98.

Formosa Oolong Tea   The rare brown-leaf tea that has the exquisite transitory aroma of ripe peaches on a sunny day. ¼ pound—$2.98.

Darjeeling Gold Tip   Nutty, classic, muscat flavor from teas grown at 6,000 feet. Picked only twice a year, brew one minute longer than others. ¼ pound—$2.98.

Zen Green Tea   From Japan's coastal plains an unfermented, basket-fired sea tea with a mystical aroma. ¼ pound—$2.98.

Orange Spice Tea   Ceylon tea combined with orange peel, cloves, and imported spices for a zesty, tangy flavor. ¼ pound—$3.98.

Oriental Jasmine Tea   Fine, flowery orange pekoe with a touch of jasmine blossoms, mild and museful. ¼ pound—$2.98.

Ceylon Breakfast Tea   Smooth full-flavored blend of Ceylon teas. ¼ pound—$2.98.

German Mint Tea   Airy, mentholated, and invigorating pickup. ¼ pound—$2.50.

Hungarian Camomile   World-famous, gentle tea, soothing and tender, one of the greatest herb teas. ¼ pound—$2.50.

See International Groceries chapter for company description.

PEDRO PINTO
Box 3208
Carmel-by-the-Sea, Calif. 93921

GUATEMALAN COFFEE

Brew the world's finest coffee right in your home. This is not a blend, just pure premium coffee grown high in the mountains of Guatemala, "Land of Highest Quality Coffee." In Guatemala, coffee plants breathe pure mountain air, and are nourished from soil rich in volcanic ash and lava. Beans from Guatemala are normally mixed with low-altitude coffees to give them a special, rich flavor.

Pedro Pinto's is the *only* coffee available in the United States today that is 100 per cent high-altitude premium coffee. You may order it preground to fit your coffee maker. Even better, order it whole bean and grind it fresh for each pot. One pound makes 36 cups of coffee. Available in 1-pound, 2½-pound, and 5-pound bags at $2.00, $4.75, and $8.75.

An ideal gift for gourmet friends, clients, and business associates.

For latest prices and ordering information write Pedro at the above address.

HEALTH AND ORGANIC FOODS

There was a time, just a little over three generations ago, when there would not have been a special section in this catalog for health and organic foods. Back in the early 1900s, nearly everything was organic. There were no preservatives, no chemical fertilizers, no mass production methods. Fresh fruits and vegetables were purchased from your local farmer, or even better, they were grown on your own land.

But now, all that has changed. In the early 1900s a German chemist, Justus von Liebig, made a simple discovery that led to a revolution in agriculture—adding chemicals to fertilizers. As a result, farmers were able to grow larger fruits and vegetables and faster than they could previously. That, combined with improved transportation and refrigeration techniques, made for an amazing change in farming methods and distribution. For years, people consumed tons of these organically impure fruits and vegetables.

The return to organic farming and natural foods started as a ground-swell revolt against the products of twentieth-century technology. The ground swell has now grown into a major revolution. People of all ages, from all over the country, are refusing to buy the overprocessed, tasteless foods with which our supermarket shelves are so abundantly stocked.

The term "organic" has become a blanket word for an entire way of life. Living organically might embrace farming naturally rich and organically fertilized soil or raising animals without hormones or other artificial growth stimulants. It can also mean leading an ecologically thrifty, nonpolluting way of life that includes eating natural foods, among other things.

Simply defined, natural foods are foods from which nothing has been taken by processing, nothing added during cultivation. Because natural or organic foods are pure, their flavor is considered to be superior. They are also thought to be healthier than their processed counterparts. People who eat health or natural foods say they feel better, are healthier, and have more energy than when they were eating foods grown and processed scientifically.

Although the health and organic food industry is growing by leaps and bounds, there are not many stores around the country that sell these foods. However, you can buy them directly from many of the growers—by mail. Just by filling out an order form and dropping it in a mailbox, you can have luscious vine-ripened fruits and vegetables; rich, creamy fresh ground peanut butter made without hydrogenated oils and dextrose; delicious bread made from stone-ground, unbleached flours. There is no need to settle for processed food when you can have the "real" thing delivered right to your door.

TIMBER CREST FARMS
4791 Dry Creek Road
Healdsburg, Calif. 95448

DRIED FRUITS

Reward yourself with dried fruit from Timber Crest Farms. Timber Crest Fruit Orchards are 100 per cent organic. Insects destructive to the fruit are counteracted by the release of predator insects rather than chemicals. Thus nature is kept in balance and the fruit allowed to mature free from defects. The ripened fruit is picked, washed in clear, cold spring water, and laid out to dry. When dried, it is inspected. Those passing inspection are placed in refrigerated storage to await customer orders. Fruit removed from storage is washed once again, honey-dipped, hand-sorted, and packaged in 5-pound bags. Apricots—$11.62. Peaches—$10.27. Pears—$7.61. Also available *bulk-dried* in 5-pound bags, not honey-dipped: Jumbo prunes—$6.11. Monukka raisins—$6.97. Apples—$8.99.

For a complete price list of bulk-dried fruits in your area, write Ronald Waltenspeil at the above address.

TROPICAL BLOSSOM HONEY CO., INC.
SUNNY SOUTH APIARIES
P.O. Box 8
Edgewater, Fla. 32032

ORANGE BLOSSOM HONEY

Romantic, fragrant, delicate orange blossoms! The flowers of young love and marriage! In early spring, before most plants bloom, thousands of acres of Florida citrus trees fill the air with their fragrance. Beehives are moved into the groves where the bees transform orange blossom nectar into honey. Packed with the beeswax combs or strained. A new, delicious flavor for honey lovers. 1-pound jar—$2.50.

TUPELO HONEY

Tupelo gum trees grow in abundance only along the riverbanks of northwestern Florida. White tupelo honey is unique. It contains about 48 per cent levulose and 24 per cent dextrose, as opposed to other honeys, which average 38 per cent levulose and 34 per cent dextrose. The high-levulose low-dextrose levels make tupelo honey safe for many diabetics. A rare honey for special people. 1-pound jar—$4.50.

For retail mail-order prices and a complete catalog, write Patricia M. Goodson at the above address.

JAFFE BROTHERS
P.O. Box 636
Valley Center, Calif. 92082

DRIED RIPE BANANAS

A unique treat for snacking, baking, cooking. Whole dried bananas with no sugar or preservatives added. This scarce red variety banana with exceptional flavor is ready to eat from the package. Try them in banana cream pies for a subtle new taste. Serve with yoghurt or sour cream, or use your imagination to develop new recipes. Ten 8-ounce packages—$3.90.

UNSULPHURED DRIED FRUITS

Organically grown, unfumigated, truly natural fruit, flavorful and nutritious. This delicious dried fruit can be served as is, softened in water overnight for compote, used in fruit soups. Or add a new flavor to pot roasts and oven roasts. And how about soaking them in rum or brandy to use in home-baked fruitcakes or flaming natural desserts? In 5-pound sizes: Peaches—$9.95. Pears—$6.95. Apples—$8.50.

For a complete catalog with latest prices and ordering information, write the above address.

APPLEYARD CORPORATION
Maple Corner
Calais, Vt. 05648

FLOUR STONE-GROUND PANCAKE MIX

This delicious preparation comes to you from the kitchen of Mary Gibson, a highly respected natural food and organic farming expert from Stockbridge, Vermont. It's made from wheat, rye, buckwheat, soy, and corn—and Mary's exact recipe is included in every bag. Make your breakfast a nutritious experience—with the flour stone-ground pancake mix. 2-pound bag—$3.00.

See Condiments, Spices, and Syrups chapter for company description.

U. S. HEALTH CLUB, INC.
Yonkers, N.Y. 10701

HIGH-PROTEIN MILK DRINK

Protein might truly be called the "building block" of which life is made. The catalyst of life—the enzymes—is protein. The hormones are protein. Your billions of red blood cells are protein. Protein is also an important source of energy, for protein is not stored in the body—you must get a new supply every day to replace all the parts of the body cells which are constantly being used up or lost.

Knowing this and wishing to carry a lightweight but concentrated source of vital protein, the 1963 Mount Everest expedition selected high-protein food supplement for their assault on the world's highest mountain.

When you mix a couple of heaping spoonfuls of High-Protein Drink into a glass of milk, water, or juice, or add it to cereal, you're assuring your body of a good extra supply of protein to help keep your physical powers up. 1 pound—$2.49.

SUNFLOWER SEEDS

Every single kernel contains twenty-two vitamins and minerals as well as protein. Sunflower seeds are nature's only ready-packed vitamin and mineral storehouse. The sunflower seed boasts significant quantities of thiamine, niacin, and vitamin D and is so rich in iron that few foods other than egg yolks and liver can compete with it. The protein content of sunflower seeds is even higher, pound for pound, than the protein in meat! And it is easily digested, highly nutritive protein. Many other valuable elements too: calcium, phosphorus, vitamin B_2, and iodine. Rich in valuable polyunsaturated oil. 14 ounces—$1.59. 5 pounds—$7.49.

WHEAT GERM COOKIES

So good the whole family will love them! They are made from whole-wheat flour with twelve times the regular amount of wheat germ! They also contain whole eggs, raw sugar, and unsaturated corn oil, plus nonfat skim milk powder, lecithin, and other good wholesome ingredients.

So delicious the children will help themselves! You can even encourage them to eat more without worrying, since these cookies are loaded with good nutrition power! 8-ounce bag—$.99. 2 bags—$1.89.

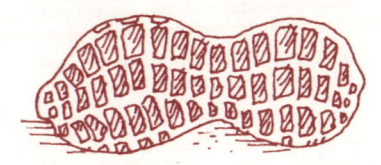

PEANUT BUTTER COOKIES

Delicious natural peanut butter cookies made with all good things, including whole-wheat flour, eggs, soy oil, and tempting pure chunky peanut butter. Everyone who loves peanuts will really go for these crispy cookies with the real nutty flavor. Kids love them . . . you will too. 8 ounces—$.99. 2 packages—$1.89.

GOLDEN SESAME COOKIES

Say "Open Sesame" to one of the most delicious cookies you've ever tasted! Golden sesame cookies are loaded with good natural ingredients like whole wheat, soy oil, sesame seeds, eggs, corn flour, sea salt, and lecithin. They're fabulously delicious and chock full of good nutritional quality. 8 ounces—$.99. 2 packages—$1.89.

CASHEW BUTTER

If you like peanut butter, you'll love cashew butter. Especially because this cashew butter contains no additives or preservatives. You get 100 per cent wholesome smooth and creamy ground cashews, with no sweetening. A most delicious nut butter. 12 ounces—$1.95.

APPLEYARD CORPORATION
Maple Corner
Calais, Vt. 05648

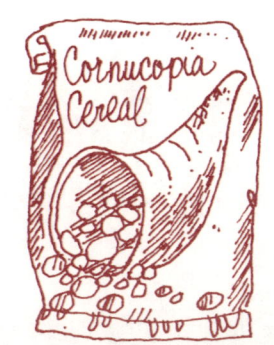

CORNUCOPIA CEREAL

For a natural nutritious breakfast, try something that's both delicious and nutritious—Cornucopia Cereal. It's made from rolled oats, Monukka raisins, chopped almonds, dried apricots, sunflower seeds, dried apples, and powdered whole orange, all skillfully blended to give you a new, wake-up taste. Cornucopia Cereal is packaged in old-fashioned coffee bags, and can also be used to bake delectable cookies and other goodies. And, of course, it contains no sugar, artificial sweeteners, or artificial preservatives. This delicious concoction is the perfect way to start your day. 13-ounce bag—$1.80. 3-pound bag—$6.00.

See Condiments, Spices, and Syrups chapter for company description.

VALLEY COVE RANCH
P.O. Box 603
Springville, Calif. 93265

ORGANIC KINNOW MANDARINS

This spring, serve freshly squeezed tangerine juice. This is the delectable juice of the Kinnow mandarin. Grown under organic conditions, you'll search a long time before you taste juice such as this. Available from April 1 to June 15, they cost $.30 per pound in 9- and 18-pound cartons.

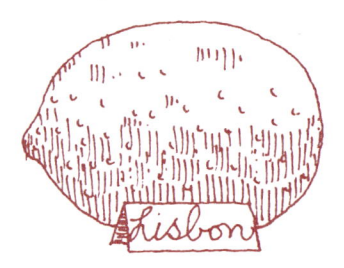

LISBON LEMONS

From December 1 to June 15, you'll be able to serve wedges of these organic lemons with all your meals. Grown under ideal conditions, they are free from chemical-type fertilizers and toxic insecticides. This means they are organic. If you've never had lemonade made with Lisbon lemons, then you're in for a real treat. They cost $.24 per pound in 9-, 18-, and 36-pound cartons.

ORGANIC SATSUMA MANDARIN

These are seedless, organic tangerines which are perfect for salads and just plain enjoyment. They are planted on the sloping, well-drained soils in beautiful Pleasant Valley (part of the Sierra Nevada foothills)—an ideal atmosphere for the production of this citrus fruit. They cost $.30 per pound in 9- and 18-pound cartons.

ABOUT THIS COMPANY
Valley Cove Ranch is located at the eastern edge of California's great San Joaquin Valley. The climate is ideally suited to the production of citrus fruit, with relatively frostless winters coupled with hot, dry summers and moderate rainfall. This ranch specializes in citrus fruits such as navel and Valencia oranges, marsh seedless grapefruits, and the fruits mentioned above. Write for their latest price list.

WALNUT ACRES
Penns Creek, Pa. 17862

BLENDED BREAD FLOUR

Even homemade bread can have a chemical-like, synthetic flavor. That's because the flour (a major ingredient) is not organically grown. Now your bread can be made the natural way. Imagine having flour that's ground fresh the day before it's shipped. This blended flour is made from organic grains which have been kept in *refrigerated* bins (unique to Walnut Acres). The flour is 100 per cent whole, entire, and complete, grown from the finest winter and spring wheats. With this flour, your bread will give you better taste—and better health. And Walnut Acres also has a fine assortment of over forty other organically grown flours and meals. You're sure to find just the flour that's exactly right for all your baked goods.

1 pound	$ .43
3 pounds	$ 1.15
5 pounds	$ 1.78
25 pounds	$ 7.30
Two 25 pounds	$13.75

See company description in this chapter.

WALNUT ACRES
Penns Creek, Pa. 17862

VEGEBURGER MIX

Turn your ordinary hamburgers into gourmet delights. And pride yourself on the fact that you've made them into an exquisite taste sensation with a healthful, organic vegetable/nut mixture. This unique combination of natural ingredients is a delicious way to spark up that old stand-by. It's made of almonds, sunflower seeds, sesame seeds, corn, oats, parsley, herbs, lemon seasoning, and other fine ingredients. Use it to add extra flavor to your backyard barbecue, picnic supper, or noon lunch. Vegeburger—for a new meal treat! 8-ounce box—$1.80.

ABOUT THIS COMPANY

For the past thirty-two years, Walnut Acres has been producing naturally grown foods—without chemical fertilizers, insecticides, or preservatives. One of the earliest of the organic farms, they have a wide selection of products, including potatoes, carrots, beets, eggs, and chickens. Write for their free catalog.

LANG'S APIARIES
8448 N.Y. Rt. 77
Gasport, N.Y. 14067

HONEY

Did you know that honey is rich in the vitamins and minerals essential to good nutrition? Do you know how many ways you can use honey? Did you know that honey retards the drying out and "staling" of baked goods? Do you know how to store honey? Have you tried pure clover honey, fallflowers honey, or buckwheat honey?

If you answered no to any of the above questions, and if you would like to taste some of the most delicious honey ever produced, Paul Lang is the person to contact. All of his honeys are thoroughly ripened in hives, handled with the greatest of care, and warranted free from adulteration. Available in 5-pound pails with recipe folders and storage hints. $6.50 per pail.

For ordering information and prices in your area, write to Paul Lang at the above address.

DIAMOND DAIRY GOAT FARM
Route 2
Portage, Wis. 53901

HEARTY SWISS CHEESE

For centuries men have known of the remarkable food properties of goat milk and have held it in highest esteem. If one goes to the oldest annals of healing, there will be found Hippocrates, known as the father of modern medicine, giving in 450 B.C. what has been called the oldest prescription, "Go to the mountains and drink goat milk." Now you can have a Hearty Swiss Cheese—*made with goat's milk*. Well known for its high mineral, vitamin, phosphorus, and protein content, goat's milk is also cholesterol-free and more easily digestible than regular cow's milk. Hearty Swiss Cheese is made from this same healthful beverage—and processed in a special low-temperature way to retain the natural flavor. Then it's aged for at least 120 days before shipment. The result is a tasty, delicious cheese excellent for weekend brunches, picnics in the park, with cocktails, or anytime. Each cheese comes in a 6½-ounce package and costs $2.00.

ABOUT THIS COMPANY

The Considines, who operate the Diamond Goat Farm, have been raising fine dairy goats for over twenty years. It is the nation's oldest organic goat dairy, offering a variety of cheese for every taste. Their goats are fed their own carefully raised and organically grown feedstuffs.

THOUSAND ISLANDS APIARIES
Clayton, N.Y. 13624

HONEY

Enjoy the wonderful flavor and fragrance of pure homemade honey—fresh from Thousand Islands own apiaries. Thousand Islands honey is processed with special care to retain its natural nutrients. And because it's prepared in small quantities, you receive the honey while it's still "ripe." This delicious natural confection is the perfect, wholesome way to satisfy your sweet tooth. You can use it on pancakes, cooked cereals, rolls, toast, muffins, ice cream, fresh fruits, pound cake . . . or even in tea. It's also just right for that "little something" Pooh Bear was so fond of. Just don't stick your head in the honey jar—you may never want to come out.

3-pound tin of liquid honey	$5.00
4½-pound tin of creamed honey	$6.50
12-ounce piece of comb honey	$2.75

WONDER NATURAL FOODS
11711 Redwood Highway
Wonder, Oreg. 97543

BEE-COLLECTED POLLEN

Pollen, the very life force of plants, is perhaps the most nutritious and concentrated food on this planet. It contains 26 per cent protein, ten essential vitamins, twelve minerals, amino acids, enzymes, and vegetable oils. Bees gather pollen dust from flowers and form small pellets held together with nectar. Only two teaspoons of these pellets each day will satisfy most adult nutritional requirements. They can be sprinkled on breakfast cereals, sandwiches, ice cream. You can blend the pollen with honey and water or fruit juice to make an Ambrosia drink. Can also be mixed with almond or cashew milk to make Ambrosia Baby Formula. And it is the perfect food for backpacking, hiking, bicycling, and traveling. $7.00 per pound postpaid.

Write the above address for ordering information.

CHICO SAN, INC.
P.O. Box 1004
Chico, Calif. 95926

RICE CAKE SANDWICH

Looking for something new to put in your child's lunchbox? Try a rice cake sandwich. This snack treat is the delicious way to have your child eat healthy food. Made from brown rice, sesame seeds, oat flour, and Chico San's own syrup, the rice cake sandwich will add healthy vitamins to every child's lunch. Or serve it for an afternoon snack. However you eat it, whenever you eat it, you'll love its scrumptious flavor and just plain good taste. 1 dozen—$3.60.

YINNIES RICE SYRUP

Here is an unusual syrup made from rice, barley, and water. This natural sweetener is free from any pollutants. Use it like honey on toast, in cooking, to flavor tea. It is known for its sustained energy. 1 gallon—$11.56.

RICE CAKES

These are low-calorie biscuit substitutes. They offer the nutrition of natural 100 per cent whole-grain brown rice and sesame seeds, without chemicals or preservatives. They come in three varieties: plain rice cakes (specify salted or unsalted), millet rice cakes (salted only), and buckwheat rice cakes (salted only). A 3½-ounce box of any cake is just $.70.

ORGANICALLY GROWN BROWN RICE

To produce an organically grown rice that will meet the rigid standards set by this company was a difficult challenge for them. After eight years of searching and talking, they found a grower in the rich Sacramento Valley of California, with naturally rich soils that had not been overcropped and had enough acres to allow a rotation plan that would build up the soil fertility rather than deplete it. The water used for irrigation of this land comes from the famous Feather River, which cascades out of the Sierra Nevada less than ten miles from this rice growing area. This pure water is uncontaminated by other nonorganic rice growers because the organic fields are the first fields to receive the water. 5 pounds of this short-grain brown rice is $3.08.

LIMA SOY SAUCE

A traditional soy sauce of exceptional quality, it is made in a little rustic factory in northern Japan by the same family for over ten generations. Chico San has visited this factory, and believes it to be the most natural and pure soy sauce that can be obtained. Lima soy sauce is aged for two years, but even after two years, enzymatic action persists; consequently, if it is not refrigerated during the hot months, white flakes may appear. This doesn't indicate spoilage or deterioration, but attests to its purity in that it does not contain chemical stabilizers nor has it been sterilized. 1 quart—$2.65.

SESAME SEEDS

These are whole brown sesame seeds. Since the outer layer is not milled from these seeds, they are very high in iron and calcium, as well as other essential minerals that are not present in the white hulled seeds. 1 pound—$1.40.

SOYBEAN PUREE

A natural preparation made from soybeans and barley or soybeans and rice or plain soybeans by the very delicate cultivation of a special natural enzyme. No chemical preservatives. When mixed with sesame butter, it makes an ideal spread on bread. Use also in soups and other cooking. 1 pound of any of the above is $1.65.

ABOUT THIS COMPANY

Chico San has been awarded the seal of certification for organically grown foods by the *Organic Gardening and Farming* magazine. Send for their free catalog.

GOLDEN ACRES ORCHARD
Route 2, Box 70
Front Royal, Va. 22630

PURE ORGANIC UNFILTERED APPLE JUICE

At long last you can obtain the pure juice of chemical-free apples. Grown in the Shenandoah Valley—perhaps the most fertile soil in America—these apple orchards are treated with organic compost along with all the trace elements known to exist. All of these nutritional elements are present in the unfiltered juice pressed from these delicious apples. Included in each shipment is literature describing the growth methods and nutritional content. 4 gallons—$8.00. Bushel of apples—$9.00.

GOLDEN ACRES ORCHARD
Route 2, Box 70
Front Royal, Va. 22630

PURE ORGANIC APPLE CIDER VINEGAR

If you've always thought of apple cider vinegar as an ordinary vinegar, you'll be surprised to learn of the *complexity* of making biologically grown apple cider vinegar. First of all, it requires the use of the *whole* apple, not just the skins (which is common to ordinary cider vinegar). Secondly, it contains a special "mother" bacteria necessary to oxidize the alcohols to acids. The resulting vinegar is ready to use in a minimum of nine to twelve months after the process has begun. No water or chemical additives used. Acclaimed by ancient people for its magic healing powers; swabbed on the wounds of Jesus because of its antiseptic properties; rubbed on the swollen and tired feet of travelers to give relief from pain and swelling; used in douches and as a mouthwash for decades because of its cleansing and antiseptic properties; used since the dawn of history in the preservation of beets, cabbage, and pickles; a requirement of all salad dressings, because of its emulsifying properties; an important constituent in weight or obesity diets because it emulsifies fats and renders them mobile; these are but only a few of the values of apple cider vinegar. A truly miraculous product. 4 gallons—$10.00.

RAINY LAKE WILD RICE CO.
P.O. Box 164
Ranier, Minn. 56668

WILD RICE

Fresh from the remote lakes and rivers of northern Minnesota comes this prized food: natural-grown wild rice. Wild rice is a nonperishable gourmet delight known throughout the world. Eighty per cent of the world's wild rice crop comes out of northern Minnesota and small amounts come out of Manitoba, Ontario, and Wisconsin. While most of the wild rice crop is sold to leading food packers and distributors, *a small portion of the crop is literally hand-selected* for special processing. This delicious rice is considered by gourmets to be the finest available in the world. After harvest, a special process parches, cleans, and grades the grains to perfection. The result is a distinctly flavorful long-grain wild rice that will complement your most prized meals and your most prized guests. Serve it in a chicken and rice casserole, with beef stroganoff as a substitute for noodles, as a side dish with fillet of flounder, or however you like it. If you think of rice as ordinary, here's your opportunity to taste a long-grain rice that's not available on your grocer's shelves. Cost of wild rice is $6.00 per pound.

VERMONT COUNTRY STORE
Weston, Vt. 05161

FRESH WHEAT GERM READY-TO-EAT

This wonderful, fresh (made weekly) wheat germ is ready to eat as a cereal, or combined with other cereals and foods to pep them up and give you a high-protein and vitamin B and E build-up. Our sealed tin can (which you keep in refrigerator) contains 1 pound of this rich, nutty, nourishing unprocessed natural wheat germ. 2 tins wheat germ—$3.50.

OTHER SOURCES FOR HEALTH AND ORGANIC FOODS

WILLIAM J. GORGUS
Route 3
Arab, Ala. 35016

This company specializes in organically grown vegetables in season.

PAUL ESKEW
4433 Escondido Cyn
Acton, Calif. 93510

This farmer specializes in beef.

BELMONT SMITH
Unity, Maine 04988

This company specializes in strawberries, vegetables, and sweet corn.

SIMONE FRUIT COMPANY
8008 West Shields Avenue
Fresno, Calif. 93705

This company specializes in organically grown fruit (dried) and nuts.

RUSSELL SMOLL
204 Hardy Drive
Tempe, Ala. 85281

This mail-order operation specializes in organically grown pecans.

OAK RIDGE ORGANIC HERB FARM
P.O. Box 1055
Alton, Ill. 62002

This company specializes in herbs.

YANKEE PEDDLER HERB & FLOWER FARM
Highway 36 North, Route 4, Box 76
Brenham, Tex. 77833

This company specializes in herb plants, cactus, flowers, vegetables, and honey.

GROW IT YOURSELF

If there is anything more satisfying than biting into a juicy piece of fruit or a fresh, crisp vegetable, it's the knowledge that you grew it yourself. And now, no one need be denied that wonderful experience. It's an easy matter for people with a little land to cultivate a vegetable patch. But, apartment dwellers, don't despair. You can now grow lettuce, tomatoes, string beans, and all sorts of other vegetables and herbs right in your own living room. All you need is a little ingenuity and the right variety of seeds.

New advances in cross-pollination have produced fruits and vegetables for every size garden, for every climate. There are tiny watermelons and cantaloupes, midget corn, "indoor" lettuce that can be grown on a window sill, along with every variety of fruit and vegetable from squash to rutabaga and even apple trees. Grow berries, tomatoes, peppers, peas, onions—almost anything at all.

But aside from vegetables, you can grow parsley, dill, basil, marjoram, and chives. As the plants mature, they can be used fresh in your cooking. Later on, you can dry them in your kitchen just as people did in colonial days.

Another highly nutritious food that everyone can grow is bean sprouts. Use a kit (see the Food Kits chapter) or the equipment already in your kitchen.

Mung, soy, and alfalfa beans, wheat berries, even lentils, are all good for in-house growing. To assure that they'll sprout, your best bet is to buy your beans from a nursery or health food mail-order firm. You'll be sure that the beans are fresh and haven't been treated with anything.

Leaf through the following pages and see how you can surprise yourself by developing a green thumb. And wait until you discover how good it feels to put out a bowl of beautiful ripe fruit and then tell everyone you grew it yourself.

FARMER SEED AND NURSERY COMPANY
Faribault, Minn. 55021

NEPAL TOMATO

A gigantic tomato from the Himalayan mountains. Very deep red in color, mid-season maturing, disease-resistant, high yielding with superb quality. Intense flavor. Really a treat for the tomato lover. Packet—$.70. ½ ounce—$5.50.

GREEN ARROW PEA

A main crop pea bred in England for the elite home and market grower trade. Disease-resistant. Nine to eleven peas in each pod. Earlier than most main crop varieties but holds well under adverse weather conditions. Excellent for freezing or canning. ½ pound—$1.85. 1 pound—$2.95.

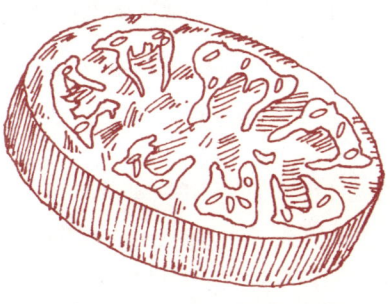

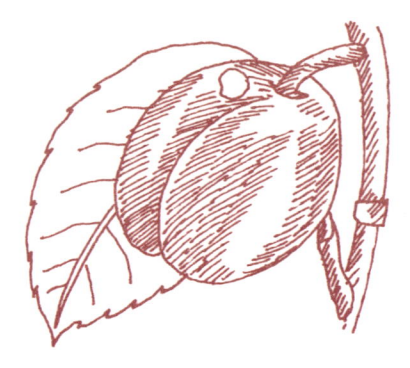

UNDERWOOD PLUM

Best known of the many plums originated at the Minnesota Fruit Breeding Farm. An annual bearer. Fruits are large, red, sweet, and juicy. Crop ripens in mid-August. Flesh is firm, pit is small. Peels readily for canning. For maximum yield, a pollinator is recommended such as toka. Trees grow 4 to 5 feet. $8.95 each.

LUSCIOUS PEARS

A high-quality dessert pear for the North developed by the South Dakota State University at Brookings. Fruit is very juicy and sweet. Has a pleasant flavor, similar to but more intense than Bartlett. Flesh is melting but firm and remains firm to the core when ripe. More resistant to fire blight than most varieties. Requires a pollinator such as patten or parker. Tree grows 4 to 5 feet. $8.40 each.

DWARF NORTH STAR CHERRIES

Super-hardy dwarf grows 5 to 7 feet high with big, glistening mahogany-red, pie cherries early in July! Morello-type, free-stone fruits make luscious pies of rich cherry-red color. Hardy throughout Minnesota, self-fertile. Trees grow 3 to 4 feet. $7.95 each.

NICHOLS GARDEN NURSERY
1190 North Pacific Highway
Albany, Oreg. 97321

A RARE BEAN FROM THE SPANIARDS OF EARLY CALIFORNIA

Santa Maria Pinquito Beans   A rare bean variety for home gardeners. They are bush type, and loaded with pods. The tiny beans are pink-colored and are used dry-shelled for baking. They are vigorous growers and may sometimes be trained as pole beans. These are a truly gourmet bean. The favored few who have tasted these beans declare the pinquitoes are the most delicious of all beans. They are low in starch and do not break up during cooking. Because of their small size and high harvesting costs, the commercial production of pinquitoes is fast coming to an end. With every order of pinquito beans there are included Portuguese baked and barbecued bean recipes. 150 seeds—$.60. 300 seeds—$1.00. 1,500—$3.00. 3,000—$5.00.

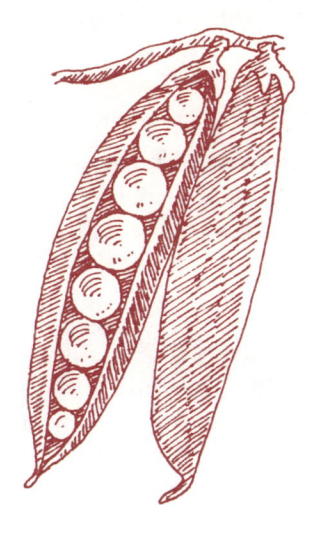

Kentucky Wonder Pole Beans   An old favorite, tasty two-purpose bean that is used green pod or dry-shelled. A heavy producer of round, thick meaty pods of superb flavor. The dried beans make an excellent vegetable protein, which can substitute for meat during present high meat prices. Packet—$.65. ½ pound—$1.65.

Christmas Limas   Pole type. A big yielder of high-quality, buttery-flavored beans. Delicious fresh, maintaining flavor and quality, canned or frozen. Beautiful carmine-striped seeds, whole color disappears during cooking. Oil the limas, adding butter, sprig of summer savory, and just before you take them off the stove, stir in a dash of heavy cream. Forget calories and start eating them, especially if you have some crusty, homemade bread to sop up the juices. Packet—$.65. ½ pound—$1.60.

Black Valentine Beans Bush type. A very old variety that has been too long neglected. A heavy producer of delicious green pods with a characteristic flavor of their own. These black beans make a lusty soup that is hard to equal. Packet—$.60. ½ pound—$1.60.

Nichols also specializes in herbs and rare seeds. Their catalog is free.

LAKELAND NURSERIES SALES
Hanover, Pa. 17331

HARDIEST PLUM TREE ON EARTH

Now for the home . . . this extraordinary Siberian plum combines the hardiness and cold climate tolerances of the European plum with the delectable, juice-laden eating of the luscious Japanese varieties! Much prized for its mouth-watering flavor, it has long flourished in the subzero regions of northern Asia and Europe. A sheer delight in spring with snowy clusters of white blossoms bursting into bloom as its green leaves unfurl. In summer it is heavily loaded with round or oval plump fruits ranging in color from a greenish yellow to red, some a deep violet. And what eating! A real taste sensation with tender flesh overrun with sweet juices! Compact and densely branched for easy, fun picking, this unusual tree reaches only 12 to 15 feet, making it a handsome addition to every landscape —large or small. $2.98 each. 2 for $5.50. 4 for $10.00.

"GIANT" BLUEBERRIES

Whoever dreamed of home-grown blueberries almost the size of a quarter? And they taste even better than they look! Blueberries from select strains like these are often whisked to America's swank hotels and cuisines by air express, so they'll arrive fresh and at the peak of perfection. Now you can pick them even fresher . . . heap bowls with their delectable, juicy goodness . . . from handsome bushes right in your own garden. For combined size, firmness, flavor, these improved strains have no peer. Plump mouth-watering berries—so big, indeed, that fifteen to eighteen berries heap a fruit dish! Make lip-smacking blueberry pies, jam— perfect for freezing too. You get a big bumper crop starting the second year after planting, and they keep producing year in and year out. 3 for $7.99. 6 for $13.99. 12 for $25.00.

THORNLESS BLACK SATIN BLACKBERRIES

These improved, thornless black satin blackberry bushes produce exceptionally high-yield crops of up to thirty huge satiny fruits per stem. They ripen early July, in bumper berry clusters, overflowing with juices and luscious honey-sweet flavor. And what a pleasure picking is . . . not a thorn to scratch you! These plants are highly disease-resistant and do not succour. They are vigorous growers with primocanes maturing to fifteen feet. This vastly superior variety has proved extremely hardy in the Midwest and South and will thrive in northern climes with winter protection. 3 for $3.49. 6 for $6.25. 12 for $11.50. 24 for $21.00.

CLIMBING TOMATOES

Now you can grow jumbo-size tomatoes in a tiny space! Only a half a dozen vines can produce bushels of big, firm fruits up to 6" across. Plump and juicy with firm meaty flesh, so delicious sliced and served fresh in mouth-watering slabs or cut up in salads all season through. You'll have loads and loads for canning, relishes, preserves, and even more to share plentifully with friends and neighbors too. You receive the kit that includes everything you need to start six vines: seeds, six nutrient-treated peat pots (with water, enlarged to full-size starter pots), tray for window sill, growing instructions. $1.59 each. 2 for $2.50.

GROW DELECTABLE FRESH MUSHROOMS

Now you can grow juicy white mushrooms right in your home! Just sprinkle a little water on the tray in this amazing kit and you and your family can not only enjoy the phenomenon of seeing it "mushroom" right before your eyes, but savor their zesty, luscious flavor in just thirty days! Oodles of the sweetest, most nutritious mushrooms you've ever tasted, and new ones will pop up each day for up to sixty days for delicious gravies, salads, garnishes! Be the envy of your friends . . . serve them steaks really "smothered" in mushrooms! Kit has growing tray, special soil, mushroom spawn plus a fascinating instruction booklet. Each kit $5.98. 2 for $11.00.

NICHOLS GARDEN NURSERY
1190 North Pacific Highway
Albany, Oreg. 97321

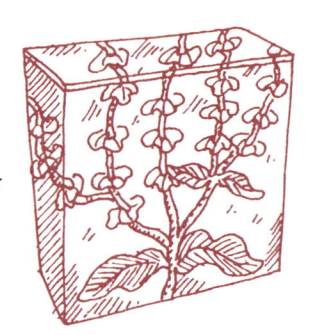

UNUSUAL HERB SEEDS

Growing herbs is a rewarding hobby that will give you many delightful hours of gardening. Unusual plants, sweet aromas, and rare flavors are the bountiest dividends awaiting the herb grower. Herbs will lead you on the road to new culinary adventures, or their medicinal properties will give you relief of some of the illnesses that plague mankind. Listed here are some unusual herbs. Herb packets contain 100 seeds, unless otherwise noted, and cost $.46 per packet.

Angelica Boil this leaf with fish. Candied stalks are great for decorating cakes, or cook these stalks with rhubarb for delicious flavor. The roots, when boiled, are supposed to give you relief of bronchitis.

Aconite This is a medicinally used herb for neuralgic pains, lumbago, and rheumatism. It is a beautiful plant.

Bene Sesame These seeds are used in breads, confectionary, and pastry. Medicinally, they are used for catarrh.

Borage These cucumber-flavored leaves are used in salads. When the flowers are sugared, they make a delicious confection. Medicinally, they are used for rheumatism.

Cardoon The root is blanched and eaten like celery. Stalks of leaves, when blanched, are stewed in soups and salads. Medically, it acts as a mild laxative.

Hyssop The young leaves are chopped fine for salad dressings. Medically, you can make tea from the leaves for coughs, colds, dyspepsia, and use it as a cathartic.

Lemon Balm The leaf can be used for tea, soup, sauces, stews, salads, and flavoring summer drinks.

Oriental Garlic Chives Grown like chives, and used like chives, except these leaves impart a slight garlic flavor to salads and other foods in which they are used. It is also a beautiful flowering plant. (25 seeds)

Purslane A pleasant salad herb, it has thick stems that, when they run to seed, are pickled in salt and vinegar for winter salads. Medically, it is very high in vitamin C and used for scorbutic trouble.

Saponaria This is a delightful flowering plant which is excellent in rock gardens. The plant produces soaplike suds when stirred in water. Many people like to use this plant to make a mild hair shampoo.

Sorrel, French This is a marvelous spinachlike leaf which is used for salads and soups. It is purple in color and high in vitamin C.

Speedwel This is a medicinal herb which is used for rheumatism, gout, and other skin diseases.

FARMER SEED AND NURSERY COMPANY
Faribault, Minn. 55021

GOLDEN MIDGET WATERMELON

Here's a sensational new melon for small gardens. A miniature watermelon. These 7″ fruits make great *individual* desserts. These rich, sweet little melons come from New Hampshire and take only sixty-five days to mature. When ripe, they turn a bright golden color, a sure indication that the fruit is at its juicy peak. An extremely early variety, they're perfect in the North. Get a jump on the season. And whenever you eat them, these Golden Midget watermelons are sure to brighten (and sweeten) up your day! Package—$.69. Ounce—$2.00.

LE JARDIN DU GOURMET
Raymond Saufroy, Imports
West Danville, Vt. 05873

SHALLOTS

If you're a gourmet cook, then you know the importance of having fresh shallots for your cooking. So why not grow your own? These delicious little gourmet onions are essential to French cooking—but they also add zesty flavor to dishes of any nationality. You'll love them in quiche Lorraine, veal Parmesan, chicken cordon bleu, beef stroganoff, shrimp with garlic, or your own favorite recipes. And these shallots are easy to plant and care for. Complete instructions are included with every shipment. 12 ounces of shallots—$2.50 postpaid.

NOTE: 12 ounces is the equivalent of 45–55 medium shallots or 25–35 large shallots.

DRIED HERBS

Dried whole leaf herbs and savory seeds can convert the common everyday foods into gourmet meals. These "natural" seasonings are full of flavor because they have not been pulverized to hide the "fillers" often found in ground spices and herbs. Herbs and spices come to you in Loc-Tite resealable polybags, to preserve their freshness. All packets contain recipe hints.

Whole Leaf Herbs
   Rosemary
      1 ounce . . . . $ .45
   Summer Savory
      1 ounce . . . . $ .55
   Sweet Marjoram
      1 ounce . . . . $ .65
   Mint
      1 ounce . . . . $ .55
   Sweet Basil
      ½ ounce . . . . $ .45

Whole Savory Seeds
   Cardamon
      ½ ounce $1.75
   Coriander
      1 ounce . . . . $ .40
   Sweet Fennel
      1 ounce . . . . $ .40
   Dill Seeds
      1 ounce . . . . $ .50

See Cheese chapter for company description.

JAMES G. GILL COMPANY, INC.
204–210 West 22nd Street
Norfolk, Va. 23517

MINIATURE COFFEE TREE

Now you can grow your own coffee at home! Believe it or not, your living room can be transformed into a small plantation—with Gill's miniature coffee tree. Grown from South American plantation seeds, this little plant is only 8–10″ tall and comes fully rooted in a plastic pot. But, best of all, it actually produces *real* coffee beans under "living room" growing conditions. Imagine how surprised your guests will be when you tell them your coffee is home-grown. And the miniature coffee tree costs only $3.00.

See Beverage chapter for company description.

OTHER SOURCES FOR GROW IT YOURSELF

MOTHER'S GENERAL STORE
P.O. Box 506
Flat Rock, N.C. 28731

See Food Kits chapter for company description.

W. ATLEE BURPEE CO.
Doylestown, Pa. 18901

KELLY BROTHERS NURSERIES, INC.
Dansville, N.Y. 14437

SPRING HILL NURSERIES
110 Elm Street
Tipp City, Ohio 45371

STARK BROTHERS NURSERIES & ORCHARDS
Louisiana, Mo. 63353

STERN'S NURSERIES
404 William Street
Geneva, N.Y. 14456

INTERNATIONAL GROCERIES

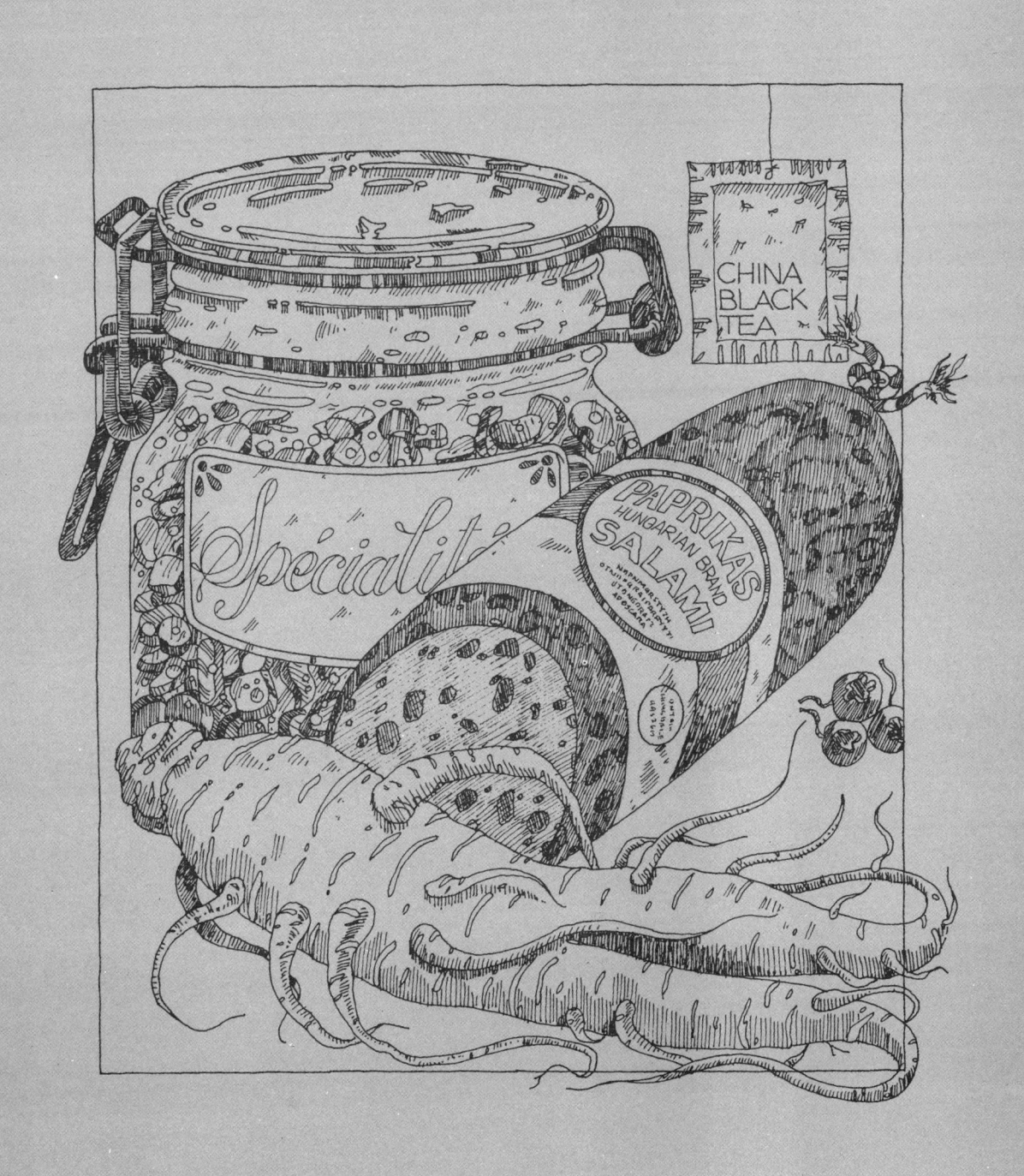

International cuisine, be it from France, India, China, Greece, or Israel, tantalizes the American palate with its variety of tastes, textures, and aromas.

For instance, in Greece, the people not only savor the taste of their food, but they delight in the *feel* of it. An excellent dish is considered to have good *yefsis,* meaning both flavor and texture. The crunch of a bite of crusty bread is as satisfying as its flavor. A juicy piece of fruit is made excellent if one can peel it without breaking the skin.

The art of eating, i.e., the art of *experiencing* food, has been nearly lost. Drive down any highway and note the proliferation of fast-food chains to see just how lost it is.

In many countries, the "flavor" of food refers not only to the taste, but to the cultural surroundings as well.

Obtaining the necessary ingredients for most international dishes would be difficult, if not impossible, if one had to rely on local stores. But, fortunately, many of the ingredients such as straw mushrooms, mustards, pita bread, cloud ears, beans, rices, and other specialized foods can now be acquired through the mail.

So whether you're an adventurous eater who wants to explore the delights of *moshari stifado* (a Greek veal dish), or a knowledgeable chef, look through the companies listed on the following pages and expand your recipe catalog with ingredients you didn't think were available. Oh, and *"Eis hygeian!"* (to your health).

MAGIC GARDEN HERB CO.
P.O. Box 332
Fairfax, Calif. 94930

GINSENG PRODUCTS

Used as a tonic in the Far East for over five thousand years, Ginseng is just being discovered in the Western world. Extensive research in Russia has shown Ginseng to prevent aging, relieve tension, fight fatigue, and restore the bodily process necessary for a healthy sex life.

    Chinese Ginseng Roots—$7.00 per ounce, $80.00 per pound
    Prince Chinese Ginseng Powder—$2.00 per ounce, $18.00 per pound
    Chinese Ginseng Extract Balls—$3.00 for 2 dozen

All prices prepaid plus postage.

MAGIC GARDEN HERB CO.
P.O. Box 332
Fairfax, Calif. 94930

CULINARY HERBS

Many cooks are still unfamiliar with culinary herbs. The art of using these aromas lies almost entirely in one word, *subtlety*. Also, even chefs are generally unaware of the vitamin and mineral contents of culinary herbs. In addition to making your food more flavorful, you can add essential nutrition.

Cumin Seeds are much used in Mexican cookery as an ingredient in chili powder. Also used in pastries, cookies, and breads. Cumin seeds contain calcium, iron, and vitamin A. $1.00 per package, $2.50 per pound.

Fenugreek Seed is used in Indian curries. $1.00 per package, $3.00 per pound.

Saffron, the most expensive spice in the world, is said to be the richest known source of vitamin B_2. Use it in breads, cakes, fancy rolls, rice dishes, soups, sauces. 15 grains for $2.00.

All prices prepaid plus postage.

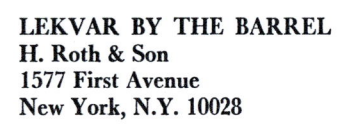

LEKVAR BY THE BARREL
H. Roth & Son
1577 First Avenue
New York, N.Y. 10028

PRUNE LEKVAR

Luscious lekvar, a very thick purée of plump sweet prunes, makes a mouth-watering filling for cakes and pastries. Keeps its smooth texture throughout the baking process . . . will not melt or run. Some people call it prune butter and use it to spread on bread for a delightful snack-treat. At Lekvar by the Barrel you can buy this ambrosial delicacy fresh from the barrel . . . in any quantity you desire. Try it soon! $1.59 per pound.

CHINA BOWL TRADING CO.
80 Fifth Avenue
New York, N.Y. 10011

BLACK MUSHROOMS (Winter Mushrooms)

These classic Oriental mushrooms add color, texture, and concentrated flavor to Chinese soups, foo yungs, noodles, vegetables, and most all stir-fried, braised, and steamed dishes. Also used in Western cookery for stuffings, gravies, and sauces and as a vegetable accent with peas or green beans. These dried mushrooms will keep indefinitely when stored in a covered jar or closed plastic bag in a dry, cool place. 1 ounce—$1.59.

CHINA BOWL TRADING CO.
80 Fifth Avenue
New York, N.Y. 10011

STRAW MUSHROOMS

These are mushrooms for soups and delicate dishes of poultry, fish, lobster, and crab that benefit from their fragrant aroma of the forest, blend of beige colors, deep bell shape, and subtle flavor. Use as a replacement for morels in Western cookery. These mushrooms will keep indefinitely when stored in a covered jar or closed plastic bag in a dry, cool place. 1 ounce—$1.69.

CHINA BOWL TRADING CO.
80 Fifth Avenue
New York, N.Y. 10011

TIGER LILY BUDS

Once you get away from "typical" Chinese cooking, one of the first ingredients you are likely to need is tiger lily buds. These delicate little dried blossoms have a slightly sweet flavor. The Orientals consider tiger lily buds to be a staple in cooking—something not to be without. They also provide an unusual way to season your other foods—homemade dishes, soups, vegetables. They're a great way to add zest to your food—any way you use them. And each 1½-ounce container has an exciting recipe printed right on the box. Each box costs only $.69.

CHINA BOWL TRADING CO.
80 Fifth Avenue
New York, N.Y. 10011

RICE STICKS

Whoever heard of Chinese spaghetti? No one, really. But here's the closest thing to it. Rice sticks are the Chinese equivalent to vermicelli. These noodles have a marvelous flavor and are used in soups, red-cooked dishes, and as a separate course with its own meat sauce. It's also used as a crunchy garnish to meals when deep-fried. Or serve it as a rice (or potato) substitute. 7-ounce package only $.99.

PAPRIKAS WEISS IMPORTER
1546 Second Avenue
New York, N.Y. 10028

EXTRAORDINARY PAPRIKA

Paprika is the very soul of Magyar cooking. It adds an exquisite and distinctively Hungarian character to meats, fish, vegetables, and salads. Paprika not only seduces the taste buds, it nips at the nostrils and enchants the eye. Contrary to popular notion, paprika is not a native Hungarian spice. It is the dried fruit of a plant indigenous to Central America and was introduced to Europe shortly after the voyages of Columbus. In Hungary, where the art of cooking is a happy marriage of East and West, its possibilities were immediately recognized, and paprika, thanks to a unique combination of soil and climate, flourished throughout the southern part of the land. Paprika, as Hungarians know it, is, therefore, different from paprika grown in any other part of the world, and the paprika grown in the region around the southern town of Szeged on the Tisza River is acknowledged as the very finest anywhere. In your refrigerator, it keeps indefinitely. If you think of paprika as just a powder to add color to dishes, you've never really enjoyed genuine Hungarian paprika. Its deep red-rose hue is a clue to its strength, and it comes in three varieties. The most widely preferred is the sweet (or mild). You may also order the half sweet (or medium) or the hot paprika (to be used sparingly). All fine cooks use paprika as frequently as salt and pepper. Add paprika to your collection of seasonings today. Sweet, half sweet, hot—$3.98 per pound.

INTERNATIONAL GROCERIES 185

ABOUT THIS COMPANY

Paprikas Weiss is perhaps the nation's largest and best known gourmet shop specializing in Middle European merchandise. For over eighty years this distinctive outlet has served Americans of Hungarian, Czechoslovakian, Yugoslavian, and Austrian descent from its location within New York's famed Yorkville section. Over 10,000 authentic foods and kitchen utensils from these foreign lands occupy every shelf and corner of the cluttered, yet folksy and highly personalized store. Visit them on your next trip to New York. An annual subscription to their catalog is $1.00 and the price is applicable to any purchase you make.

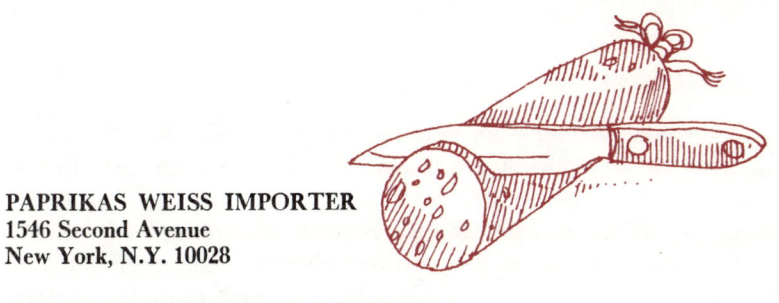

PAPRIKAS WEISS IMPORTER
1546 Second Avenue
New York, N.Y. 10028

HUNGARIAN-BRAND SALAMI

For Hungarians the world over, pleasant memories of the homeland include the sweet experience of Magyar cuisine—a feast of color to delight the eye, a bouquet of heady aromas to spur the appetite, a medley of subtle, succulent flavors. This has been re-created in Paprikas Hungarian-brand salami. The ingredients are a well-guarded family secret, but it is made only during the cold months of the year. That's why Hungarians call it *teli szalami,* or winter salami. It is cured for months, a slow drying-out process with temperature and humidity carefully controlled to create its distinctive, delicate flavor. Keep it in your refrigerator and it will remain fresh and piquant for a long, long time. Take it on picnics, serve it will remain fresh and piquant for a long, long time. Take it on picnics, serve it However you like it, it's likely to be the best salami you've ever tasted. The Hungarian-brand salami comes in 1-to-5-pound sticks. Cost is $5.98 per pound.

See company description in this chapter.

PAPRIKAS WEISS IMPORTER
1546 Second Avenue
New York, N.Y. 10028

DOBOSH TORTE—THE ORIGINAL VIENNESE SEVEN-LAYER CAKE

Imagine seven thin layers of butter-rich, fine-textured cake between layers of rich, creamy chocolate! Imagine a thick icing of luxury chocolate on three sides!

This exquisite dessert is a Continental tradition you can enjoy anytime in your own home. Made of the highest quality ingredients, it will keep as long as three months in your refrigerator or freezer. Approximately sixteen servings per cake. Foil gift-wrapped. 18-ounce box—$4.98. 3 boxes—$14.00.

See company description in this chapter.

PAPRIKAS WEISS IMPORTER
1546 Second Avenue
New York, N.Y. 10028

STRUDEL DOUGH SHEETS

Maybe you remember the strudel your grandmother used to make. The house was filled with fragrant, tantalizing aromas all day long, as her deft fingers mixed and shaped the dough into paper-thin, flaky, almost transparent sheets and spread on a variety of mouth-watering fillings. Strudel is a triumph of the confectioner's art. Pastry shops have flourished in Europe's great cities, on the quality of their strudel alone. Famous Continental chefs have won international acclaim based on their skill in baking strudel. The secret lies in the texture and consistency of the dough. Now, even if you've never made a strudel before, this fresh strudel dough practically guarantees success. This time-tested blend of ingredients is fast and easy to use. The illustrated six-step instructions plus recipes for fillings are in every package. Just fill the sheets, bake for 30 minutes, and enjoy! 1 box (4 sheets)—$.98. 12 boxes—$11.00.

See company description in this chapter.

PAPRIKAS WEISS IMPORTER
1546 Second Avenue
New York, N.Y. 10028

THE STORY OF MONT BLANC CHESTNUTS

For Europeans, chestnuts and Christmas are synonymous. All over the Continent, children sniff the aroma of chestnuts in the kitchen and know that holiday festivities and dreamed-of joys are nearly here. If you were ever a Yuletide dinner guest in Paris, Vienna, or Budapest, chances are the most memorable course you were served was this traditional dessert. The French call it *Mont Blanc,* the Viennese

INTERNATIONAL GROCERIES 187

Kastanienreis, and Hungarians know it as *Gesztenyepure,* or riced chestnuts. *Gourmet* magazine paid tribute to these chestnuts. You can serve this international delight in your own home at Christmas or any time of the year. Just open a can of Veron purée of chestnuts and pass it through a Mont Blanc ricer to make it fluffy. Sprinkle lightly with rum. Top with fresh whipped cream. Chill and serve. *Voilà!* Your reputation as a great cook is made. 8-ounce tin—$1.25. 3 tins—$3.50. 6 tins—$6.75. Case of 48—$52.00.

Mont Blanc chestnut ricer. The secret of success! Handy gadget makes chestnut purée light, fluffy. Easy to operate. Many more uses. A "something different" gift. Triple-plated chrome and enamel finish. 11-inch length; 3½-inch cup diameter. Each $3.98. 3 for $11.00.

See company description in this chapter.

PAPRIKAS WEISS IMPORTER
1546 Second Avenue
New York, N.Y. 10028

THE GREAT TEAS OF CHINA

At last tea lovers can enjoy the many excellent varieties of tea from the land where it all began. Tea was grown in China in prehistoric times. Through the centuries, the Chinese have patiently cultivated their tea plants, bred and crossbred them to produce many new strains, turned the preparation of tea leaves into a fine art and the brewing and serving of tea into a ceremony that is a celebration of gustatory delights. Now, after a lapse of many years, you can have all the pleasures of sipping genuine Chinese tea, grown in Mainland China, handpicked, processed, and packed in Mainland China. Discover a world of new taste experiences awaiting you with this first shipment of famous Chinese teas. Each in its own way is a tea to tease your palate with delicate nuances of flavor. These exquisite Chinese teas, each in a beautiful canister, are too lovely to keep on your kitchen shelf. Each contains 8 ounces of premium Chinese tea. The perfect gift for your tea-sipping friends. Lung Ching Green Dragon Tea, a very long-grained, velvet-textured green tea from Hangchow, the pleasure city of China. Keemun Black Tea, famous for its fruity bouquet, a rare treasure among teas. 8-ounce canister—$4.98. 3 for $13.00. 6 for $25.00.

See company description in this chapter.

PAPRIKAS WEISS IMPORTER
1546 Second Avenue
New York, N.Y. 10028

HUNGARIAN SAUSAGE

In Europe, three generations of hikers, mountain climbers, cross-country skiers, hunters, campers, and Boy Scouts have made Paprikas Hungarian-brand sausage their standard fare. No other food has such highly concentrated nutritional value. A week's supply fits easily in a small corner of a knapsack. Well cured and sold in heat-sealed cellophane wrapping, a Paprikas Hungarian-brand sausage will stay fresh and savory indefinitely. These sausages are still made by Hungarian experts, following a time-tested recipe, using only pure pork meat and sweet Hungarian paprika for pungency. Months of patient curing gives them a very mild, smoky flavor that is rich and satisfying. But you don't have to be a mountain climber to enjoy this Old World delicacy. Slice it thin and serve with toothpicks at your next cocktail party. Arrange on a well-garnished platter as a special buffet treat. 1 pound—$5.98. 5 pounds—$28.00. 10 pounds—$54.00.

See company description in this chapter.

PAPRIKAS WEISS IMPORTER
1546 Second Avenue
New York, N.Y. 10028

IMPORTED WILD FRUIT SYRUPS

Pure raspberry syrup, from the fresh fruit, with no artificial coloring. Among the finest quality made. Unlike some syrups, it is not overly sweet. Its snappy flavor has a kick to it and the aroma is a pleasure in itself. In Vienna's many cafés, it is still enjoyed with a "spritz" of club soda as a refreshing summer drink. You, too, can enjoy it as a tangy, nonalcoholic beverage at cocktail time. Use it also as a topping for puddings and cakes. Serve with sliced peaches on ice cream and *voilà!* . . . peach Melba.

Pure Wild Raspberry Syrup   24-ounce bottle—$2.59. 3 for $7.50. 6 for $14.00. Also comes in 11½-ounce bottle—$1.79. 3 for $5.00. 6 for $9.50. The following flavors come in 11½-ounce bottles—$1.79; 3 for $5.00; 6 for $9.50; 12 for $18.00: **Wild Strawberry Syrup, Wild Red Currant Syrup, Wild Black Currant Syrup—Cassis, Wild Sour Cherry Syrup, Wild Elderberry Syrup, Wild Fruit of Rose—Hagebutten,** and **Wild Gooseberry.**

See company description in this chapter.

INTERNATIONAL GROCERIES

PAPRIKAS WEISS IMPORTER
1546 Second Avenue
New York, N.Y. 10028

IMPORTED ZAGREB HAM

The ham of Zagreb, mildly cured and cooked, Yugoslav style, to exquisite succulence, is prized throughout the Balkans. And everyone in that part of the world knows that no better ham is available than the boneless and skinless ham with the famous Gavrilovic label. Now you can enjoy its tantalizing flavor anytime you wish. Slice it right down from one end to the other. Serve it cold in sandwiches or as hors d'oeuvres or heat it for a family dinner. 1-pound tin—$2.95. 3 for $7.00. 6 for $13.00.

See company description in this chapter.

MAISON GLASS
52 East 58th Street
New York, N.Y. 10022

STRASBOURG FOIE GRAS WITH TRUFFLES

For centuries Strasbourg has been the goose-liver capital of the world. The finest quality pâté and goose-liver products are only those actually packed in Strasbourg. Order Délices de Strasbourg Foie Gras laced with rare and delicate bits of truffle. You and your guests will taste the difference. 5¼-ounce tin—$16.95. 7⅜-ounce tin—$23.95.

MAISON GLASS
52 East 58th Street
New York, N.Y. 10022

DINNER FOR TWO

Make a picnic supper really special with imported gourmet foods. Start your repast with *pâté de foie de Strasbourg*, then *potage de légumes*. Your entree will be smoked pheasant served with wild rice and *petit pois*. For dessert—*baba au rhum*. Sheer elegance! Packed in a handsome, reusable wicker hamper. $35.95 complete.

OTHER SOURCES FOR INTERNATIONAL GROCERIES

STAR MARKET
3349 North Clark Street
Chicago, Ill. 60657

This company specializes in Chinese food.

BEZJIAN GROCERIES
4725 Santa Monica Boulevard
Hollywood, Calif. 90029

This company specializes in Indian food and has a $10.00 minimum order.

INDIA GROCERS
5002 North Sheridan Road
Chicago, Ill. 60640

This company specializes in Indian foods.

ORIENTAL IMPORT-EXPORT COMPANY
2009 Polk Street
Houston, Tex. 77002

This company specializes in Chinese food and has a $12.00 minimum order.

CASSO BROTHERS
570 Ninth Avenue
New York, N.Y. 10036

This company specializes in Middle East foods and has a $25.00 minimum order.

AMERICAN ORIENTAL GROCERY
20736 Lahser Road
Southfield, Mich. 48075

This company specializes in Chinese and Middle Eastern foods.

SAHADI IMPORTING COMPANY
187 Atlantic Avenue
Brooklyn, N.Y. 11201

This company specializes in Indian foods and has a $15.00 minimum order.

SHING CHONG & COMPANY
800 Grand Avenue
San Francisco, Calif. 94108

This company specializes in Chinese foods.

SPECIAL FOODS

Just as mankind has always been anxious to explore new and uncharted frontiers, so has he also been on the prowl for new ways to vary his diet. Whether it be for health reasons or just for variety, special foods have always intrigued us. And through the years, the selection of foods has broadened to satisfy nearly every culinary curiosity.

Today, more than ever, people are intensely interested in food and creative ways to serve it. (Cookbooks are the largest selling category of nonfiction books in the United States.) Consequently, the adventurous are constantly searching the globe for new and unusual dishes. Now, you need not go further than your own living room. There is a whole world of special foods available by mail order.

Imagine ordering your next pizza from a mail-order catalog, or unveiling, at your most elegant dinner party, a delicious Crêpe à la Reine (which has been delivered by your postman just that morning). Now you can buy everything from diet foods and survival kits to caviar and antipasto through the mail.

Or, if you are a camping enthusiast, you'll be equally pleased with what's new. You may camp as a means of having a quiet and isolated experience, or to get in touch with the pioneer spirit. Whatever you desire, you'll be pleased to know that you don't have to give up your gourmet eating habits while on your adventure. Buying provisions for a camping trip can be quite a problem. But not anymore. You'll have a large selection from which to choose, and you'll be saving yourself the effort and bother of going from store to store for individual (and hard-to-find) items.

So if you, or someone you know, are interested in special foods, read through the following listings carefully. These companies produce all conceivable types of foods. And the U. S. Postal and UPS can bring these marvelous substitutes right to your door, adding new zest and pleasure to what might otherwise be a monotonous diet.

RECREATIONAL EQUIPMENT, INC.
1525 Eleventh Avenue
Seattle, Wash. 98122

COMPLETE MEALS

If you're adventuresome and like to spend time in the great outdoors, you'll love these freeze-dried convenient foods.

The freeze-dry process of preserving and dehydrating food has progressed so much that dehydrated food is tastier and lighter in weight than ever before. Recreational Equipment offers a large selection of meat, vegetables, desserts, drinks, and even complete meals in freeze-dried or dehydrated form. **Breakfast #1**—instant applesauce, pancake mix, maple syrup mix, cocoa, cooking oil. $3.00.

Breakfast #2—orange juice, eggs 'n' Bac-o-Bits, hash brown potatoes, cocoa, cooking oil. $3.75. **Lunch** #1—beef jerky, Swiss cheese/imitation bacon bits, crisp toast, grape drink. $3.50. **Quick Trail Lunch** #6—beef jerky, cookies, 4 chocolate bars, 4 quick-energy fruit bars. $4.60. **Quick Trail Lunch** #7—natural nuggets, apple chips, 4 beef sticks, lemon and lime gatorade. $4.30. **Dinner** #1—chicken rice soup, chili mac with beef, chocolate pudding, fruit punch. $4.60. **Dinner** #2—vegetable beef soup, quick rice/chicken, French apple compote, lemon-lime drink. $5.10. **Dinner** #3—tomato noodle soup, vegetable beef stew, dumplings, butterscotch pudding, orange drink. $5.65. **Dinner** #6—chicken noodle soup, stroganoff beef, blueberry cobbler, lemon-lime drink. $5.70.

All contain four servings per package.

OMAHA STEAKS INTERNATIONAL
4400 South 96th Street
Omaha, Nebr. 68127

CRÊPE À LA REINE

Try a memorable delicacy—a thin crêpe filled with fresh mushrooms and diced chicken, and sautéed in a cream sauce with a touch of sherry. Simply place on a baking sheet 20 minutes at 325° and serve to fortunate guests. Then top it with your favorite sauce. 24 (2½-ounce) portions—$24.50.

CANNELLONI À LA MILANESE

Treat yourself and guests to a very fine crêpe filled with seasoned pork, veal, chicken, mushrooms, and cheese. Two portions make a serving, but you may prefer to serve one as a truly sensational appetizer. Complement these marvelous crêpes with your own seasoned sauce. 24 (2½-ounce) portions—$23.50.

HORS D'OEUVRES

Win oh-h-h's and ah-h-h's with this delightful assortment. 16 each of five party pleasers: beef and mushroom turnovers, frankfurters in blankets, mushroom caps stuffed with beef, egg rolls with chicken and mushrooms, and potato knishes. 80 treats—$28.50. 160 treats—$45.00.

SPECIAL FOODS

ABOUT THIS COMPANY

This company has one of the finest reputations in the business. Fred Simon meticulously oversees the mail-order operation. Their guarantee is a good one, and their company stands behind it.

RECREATIONAL EQUIPMENT, INC.
1525 Eleventh Avenue
Seattle, Wash. 98122

INSTANT FREEZE-DRIED CASSEROLES

Did you ever wonder what to serve your family or gourmet friends on that long camping trip? Here's a unique idea: freeze-dried casseroles. These delicious main course dinners are packaged in aluminum serving containers; just add hot water and they are instantly ready to eat. You'll find these easy-to-fix meals so much tastier—and more lightweight—than ever before, because they're processed by the special new freeze-dried method. And you can choose a one- or two-serving size in four varieties: beef amandine, turkey Tetrazzini, chunk chicken with carrots, and tuna à la Neptune. Next time you're packing for that camping trip, why not take a gourmet meal along? Tea Kettle Freeze-dried Casseroles are $.80 each for a single serving, $1.92 for a double serving.

Recreational Equipment, Inc., also produces a wide line of other camping foods and equipment. For a complete catalog, write the above address.

PERMA-PAK
40 East 2430 South
P.O. Box 15695
Salt Lake City, Utah 84115

FREEZE-DRIED MEAT

Who ever thought camping could offer so many food luxuries with such simplicity? Breakfast, lunch, and dinner can all be interesting, yet easy to do. In fact, many breakfasts and lunches offer a "no-cook" feature for early morning starts and on the trail activity. Choose from **Breakfast #1**, oatmeal 'n' apple with creamer, fruit galaxy, and cocoa. Or, **Breakfast #4,** omelet and hash browns, apple-raspberry fruit, cooking oil, cocoa. Or, **Lunch #5,** peach preserves, Yukon biscuits, tropical chocolate bars, trail snack, and drink mix. Or, **Lunch #6,** apple preserves, Yukon biscuits, beef jerky, and vanilla shake. Or, **Dinner #2,** mountain stew, skillet bread, lemon pudding, cooking oil, and drink mix. Or, **Dinner #5,** spaghetti 'n' sauce, peas and carrots, vanilla pudding, and drink mix. Or, **Dinner #8,** chicken-style stew, dumplings, chocolate pudding, and drink mix. Meticulously prepared to meet Perma-Pak's exacting standards, these four-person main-dish meals are easy to fix and fun to eat. And the meals which require cooking are packed in burnable containers, which can be easily disposed of without littering.

Breakfast #1	$4.19
Breakfast #4	$5.60
Lunch #5	$5.09
Lunch #6	$6.91
Dinner #2	$5.25
Dinner #5	$4.72
Dinner #8	$4.25

PEPPERIDGE FARM
Old Post Road
P.O. Box 119
Clinton, Conn. 06413

BREAKFAST GIFT ASSORTMENTS

Remember those hearty, leisurely breakfasts when the luscious aroma of pancakes and bacon filled the house—and there was nothing to do but enjoy it! It's still a wonderful way to start a day—or a weekend. Midnight breakfasts are fun, too. Now you can have a special selection of breakfast goodies which includes: 12 ounces of hickory-smoked bacon; 16 ounces of Grade A Vermont maple syrup; 3 bags (1 pound each) of regular, whole wheat, and buttermilk pancake mixes; 10 ounces of strawberry fruit spread; 8 ounces of Dutch process cocoa; 12 ounces of blended honey; 1 pound of corn pancake mix; 12 ounces of Open Kettle sugar cane syrup; 1 pound of Colombian supreme coffee; and a lovely Bennington pottery pitcher for syrup or cream (holds 8 ounces). $29.95.

See Confections chapter for company description.

SPECIAL FOODS

AMERICAN STORABLE FOODS
13 Arcadia Road
Old Greenwich, Conn. 06870

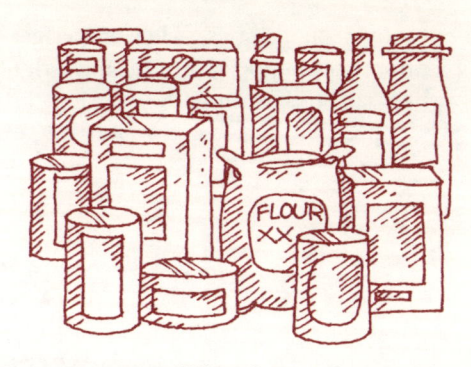

SURVIVE FOR A YEAR!

Do you know why Mormons appear so serene? They know that they and their families will eat—no matter what. Why? Because Mormons have made it a point of their belief since 1820 to have at least a year's supply of food stored for every family member. Rainy Day Foodpaks have been developed by Mormons, the foremost experts in food storage techniques. But you don't have to have religion to store food. If you would like to have a year's supply of food, for $398.15, here's what you get: corn meal, rolled oats, white rice, fruit and nut cereal, spaghetti and macaroni, barley, wheat, pinto beans and red beans, split peas, whole green peas, navy beans, soy beans, textured vegetable protein, onions, cabbage, green beans, tomato crystals, mixed vegetables, carrots, potato granules, gelatin, applesauce, prunes, banana slices, apple slices, date dices, instant nonfat dry milk, shortening powder, beef and chicken bouillon, peanut butter powder, cheese powder, scrambling egg mix, sprouting seeds, tray, sprout handbook, cookbook on sprouts, plastic lids, etc. It may not make gourmet meals, but it may just be the only meal in town.

ABOUT THIS COMPANY
Built around a modern six-acre plant, Rainy Day does all its own quality production using equipment specifically designed for packaging long-term storage food items. The heart of this operation is NiVac, which encloses food in a nitrogen atmosphere for a long shelf life. The absence of free oxygen in each container greatly reduces the possibility of any food oxidation, one of the main causes of spoilage.

STAFFORD N. GREEN
P.O. Box 625
Charleston, S.C. 29402

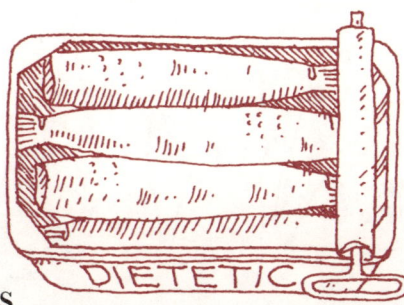

DIETETIC SARDINES

Here's the ideal food for weight watchers: "Crossed Fish" dietetic brisling sardines. These delicious, nutritious little morsels of fish are packed *without salt or oil*—to cut down on calories. They're perfect for canapés, hot casseroles, cold salads, picnics, quick suppers, or as a snack right out of the can. However you eat them, brisling sardines are mouth-watering taste treats. Price for a carton of 25 cans is $13.95 postpaid. Each can weighs 3¾ ounces.

MANGANARO FOODS
488 Ninth Avenue
New York, N.Y. 10018

MILANESE-STYLE ANTIPASTO

A delicacy as served in finest Milanese homes; packed in gift-styled jar. Contains sardines, tuna fish, mackerel, artichokes, stuffed olives, capers, and vegetables in piquant sauce of tomato paste. From Milano. 6-ounce jar, Fra-Polli—$2.50.

ABOUT THIS COMPANY

Manganaro's was founded in 1893, in its present location. It is now in its third generation of family ownership. Specializing in the "difficult to obtain," their buyers visit Italy several times a year to seek out new products, thus maintaining their reputation as the finest Italian food store in the United States.

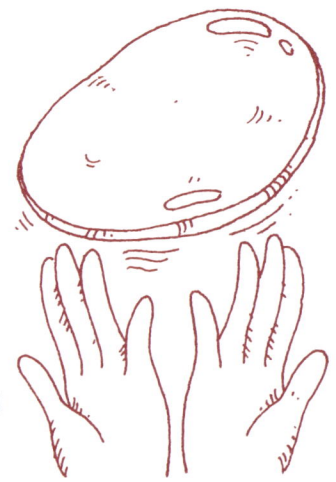

PFAELZER BROTHERS
4501 West District Boulevard
Chicago, Ill. 60632

PFAELZER PIZZA

Here's the finest pizza available anywhere—made especially delicious with Pfaelzer's own recipe. This mouth-watering entree is made of the best ingredients

SPECIAL FOODS

money can buy—San Joaquin Valley tomatoes, Parmesan, Romano, and asiago cheeses, generous portions of shredded mozzarella, and a super-thin crust—then it's skillfully baked to a golden brown. The result is a fantastically flavorful feast. And the Pfaelzer Pizza comes in three exciting varieties: sausage, pepperoni, and party (with sausage, pepperoni, and red and green peppers).

Six 12″ sausage pizzas	$20.00
Six 12″ pepperoni pizzas	$18.00
Six 12″ party pizzas	$22.00

See Meat, Fish and Poultry chapter for company description.

PORILOFF EPICUREAN IMPORTS
PUREPAK FOODS, INC.
542 La Guardia Place
New York, N.Y. 10012

ALL YOU EVER WANTED TO KNOW ABOUT CAVIAR

Caviar originated in ancient days when fishermen cured the eggs of sturgeon from the Caspian Sea with salt, to be used as a food product. The delicate and delicious taste of caviar became very popular and a tremendous demand was soon created for it. Later, as transportation developed, caviar became the world's greatest delicacy.

The following fish have eggs which are used to prepare caviar: sturgeon, salmon, whitefish, and lumpfish. The method of curing sturgeon caviar is determined by the maturity of the eggs when they are removed from the fish. If the eggs are at the right degree of maturity, they can be cured with very little salt and will be known as malossol caviar (*malo:* little; *sol:* salt). If the eggs are not fully mature, they are cured as kegged caviar with regular salt. If the eggs are overmature, they are too soft to be cured as grain caviar. Therefore, the eggs are pressed in a linen bag, giving us pressed caviar. Pressed caviar requires a very small amount of salt. No caviar is completely free of salt.

Sturgeon Caviar   Today, the best caviar still comes from the Caspian Sea sturgeon. Both Iran and Russia border on the Caspian Sea, so there is no difference between Iranian and Russian caviar. The natural color of all sturgeon caviar varies from gray black to black. Sturgeon caviar is always sold in its natural color; coloring is never added. Poriloff Green Label caviar is available in gift boxes of two 2-ounce jars, $10.90, and four 2-ounce jars, $20.75.

Salmon or Red Caviar Salmon caviar is the eggs of salmon cured with salt. Because salmon eggs have a natural red or orange color, salmon caviar is known as red caviar. Some of this caviar is packed as natural color, but in most cases, coloring is added to make the caviar a deeper red. Most salmon caviar comes from Alaska and Washington State. Salmon caviar is both nutritious and delicious. It makes wonderful hors d'oeuvres and is also used in dips, with sour cream, and in other mixtures. Four 2-ounce jars—$14.75.

Lumpfish Caviar Lumpfish caviar is the eggs of lumpfish cured with salt. This caviar is imported from Iceland. The natural color of lumpfish eggs is tan; therefore black coloring is always added to this caviar. Four 2-ounce jars—$10.95.

Whitefish Caviar Whitefish caviar is the eggs of whitefish cured with salt. This caviar comes from a vast area of Canada and the United States in the vicinity of the Great Lakes. Whitefish caviar is therefore imported from Canada or it is domestic. In either case, it looks and tastes the same. Whitefish eggs also have a natural tan color, and black coloring is always added. Four 2-ounce jars—$9.95.

SMITHFIELD HAM AND PRODUCTS CO., INC.
Smithfield 8, Va. 23430

SMITHFIELD SPORTSMAN'S BOX

Please the outdoor man or woman with this hearty collection of James River Smithfield convenience foods. Designed by sportsmen for sportsmen, the handy take-along box contains 12 cans of delicious foods, ready to heat and eat. Three 7½-ounce cans each of James River Smithfield chicken Brunswick stew, beef

SPECIAL FOODS 201

stew, and chili con carne with beans; one 10-ounce can of James River Smithfield pork barbecue; one 10½-ounce can of James River Smithfield beef barbecue; and one 5-ounce can of James River Smithfield turkey barbecue with Smithfield ham. $10.95 prepaid.

RAINY DAY FOODS
P.O. Box 71
Provo, Utah 84601

LOW-MOISTURE FOODS

"It wasn't raining when Noah built the ark"—that's why Rainy Day Foods created this unique food insurance program. Packaged in heavy-gauge metal cans with special enamel lining inside and out to prevent corrosion are foods like applesauce, cottage cheese, tuna fish, and split peas. These alone are enough to make several delicious lunches, even after fifteen to twenty years! They weight 90 per cent less than regular canned foods and can be stored in 50 per cent less space. Simply add water to re-create their fresh, natural flavors.

4 pounds applesauce	$ 9.40
1 pound 2 ounces tuna fish	$22.80
1 pound 2 ounces cottage cheese	$10.85
5 pounds 8 ounces split green peas	$ 3.75

Rainy Day Foods also offers over 150 other varieties of foods, including fruits, vegetables, grains, cereals, soups, dairy products, meat substitutes, desserts, and fruit drinks. If you wish to keep your cupboard well stocked as a hedge against inflation, unemployment, or a national emergency; or if you want to prepare for a long camping trip, write the above address.

GREAT VALLEY MILLS
Route 309
Quakertown, Pa. 18951

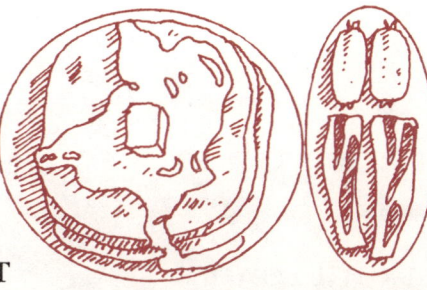

FARMER'S BREAKFAST

If you're a breakfast lover, and looking for the unusual, stop right here. Now you can purchase an unusual combination of items. Over 2½-pound slab of triple-smoked Pennsylvania Dutch bacon, 1 pound of a delicious genuine farm-made country sausage, ½ pint of pure Vermont maple syrup, three 1-pound packages of assorted stone-ground pancake and waffle mixes, and one jar of preserves are all packaged together to make a marvelous breakfast treat. $14.95 delivered.

ABOUT THIS COMPANY
The Great Valley Mills, in Bucks County, carries on a tradition born in 1710 when buhrstones were used to grind grain for flour at their original site. This concern, which furnished flour for the Continental Army during their encampment at nearby Valley Forge, is probably the oldest continuous business in Pennsylvania.

ESTEE CANDY CO.
169 Lackawanna Avenue
Parsippany, N.J. 07054

DIETETIC LEMON SANDWICH COOKIES

If you miss desserts because you're on a special salt-free diet, here's a treat for you: dietetic lemon sandwich cookies. These delicious light cookies are made with two lemon-flavored wafers, separated by a layer of cream filling. There's no salt added to these tasty pastries, so you can feel free to eat as many as you like. Try them with lemon sherbet, as something to nibble on with your afternoon tea, or as a handy snack anytime. Whenever you have them, however you have them, you'll find they're an easy way to stick to your diet. A 5-ounce box costs just $.89.

Estee also produces a fine line of other dietetic products, including bubble gum, chocolate chip cookies, hard candies, gum drops, and chocolate bars. For free catalog, write the above address.

WESTERN DIETARY PRODUCTS CO.
P.O. Box 552
Bellevue, Wash. 98009

VEGETABLE BROTH POWDER

Want to make a salt-free soup? It's easy with Vegetable Broth Powder. This amazing concoction is made entirely of dried, powdered, uncooked vegetables—

which spring to life under the influence of boiling water. Just one teaspoon of Vegetable Broth Powder and one teacup of water are all you need for a delicious soup. Try it the next time you want something different for lunch, appetizer, or late-evening snack. It's the healthful way to eat.

> 8 ounces .... $ 4.00
> 1 pound .... $ 7.50
> 5 pounds.................................... $33.00

FOOD KITS

The twentieth century has been an era where Americans have welcomed, with open arms, prepackaged, preprocessed, even precooked foods. But the art of "starting from scratch" has not been lost. It survives in families that care about what goes into their foods. And families that pass favorite "scratch" recipes from generation to generation know how much fun food making is for the entire household.

But did you know there are food kits that involve practically every aspect of eating, and let you develop your own family recipes? Yoghurt makers let you create your own flavors; jelly makers bring back that wonderful aroma from the good old days; and pickling kits, sausage makers, sourdough bread kits, lollipop kits, brandied fruit makers, cheese makers, beer and wine makers, all give you an endless selection of things to make. (One important reminder about making wine at home: If you really start liking the wine you make, we suggest that you visit the nearest Federal Alcohol and Tobacco Tax office. You are required to file Form 1541. It allows you to legally make up to two hundred gallons of wine a year for your own consumption.)

Although some of the kits reviewed in this chapter require that you make an initial investment, not only will you have the capability to produce delicious food, but in the long run, you will save money. Since you can use the equipment over and over, the cost per batch will drop. And as you become more experienced, you are likely to be able to create a greater variety of foods. Once you start making food from "scratch," it will be difficult to convert your life back to "supermarket" style.

CHRIS-CRAFT
Algonac, Mich. 48001

HOMEMADE ICE CREAM KIT

Now you can whip up fresh, pure homemade ice cream without moving a muscle. This machine makes a full quart of the richest creamiest ice cream (or ice milk, if you're watching your calories) without the corrosive mess of rock salts and ice, and no hand churning. Just pour your favorite natural ice cream ingredient into the ice cream machine, put the whole thing into your freezer, and plug in the extra-long 7½-foot cord. In an hour you'll have a quart of tasty natural ice cream with a texture and clarity of flavor that can't be bought. No artificial colorings, flavorings, or preservatives. 8″ diameter and 8″ high, compact enough for freezer compartments of many boat or wreck vehicle refrigerators, as well as for home. Choice of two delicious colors, strawberry or chocolate. 110 volt AC only. With recipe book. $24.95.

HORCHOW COLLECTION
P.O. Box 34257
Dallas, Tex. 75234

HOMEMADE MAYONNAISE

Here is an easy way to make a pint of delicious fresh mayonnaise. And since mayonnaise is such an integral part of many fine gourmet sauces, you'll find it a must. The clear-glass jar has recipe and simple instructions printed right on it in yellow. Comes with aluminum lid and plunger plus a storage cap. $10.00.

WELCH FOODS, INC.
Specialty Products Division
2 South Portage
Westfield, N.Y. 14787

THE HOME JELLY-MAKING KIT AND WINE-MAKING KIT

Welch's, the most experienced and respected name when it comes to grapes, now offers you two do-it-yourself kits to give you the pleasure of making homemade jelly and wine, too! There's a special magic about homemade jelly. The irresistible smell, the smiling faces of children waiting to lick the spoon, and that special satisfaction in knowing you did it all yourself . . . with a little help from Welch's. This kit contains all the necessary supplies for twelve jars of homemade grape jelly, including tumblers, recipe, and directions, for $5.99. Grapes are the basis of the finest traditional wines. And who knows more about grapes than Welch's? Start your home winery with one of Welch's wine-making kits. Each kit includes all the required ingredients, supplies, and step-by-step instructions you need to become your own wine maker. You can choose from red burgundy, white chablis, and rosé. $12.99.

ABOUT THIS COMPANY

Everyone has heard of Welch's. Since 1869, the Welch name has been respected for quality, excellence, and good taste. When Dr. Thomas Bramwell Welch first applied the theories of Louis Pasteur to processing grapes, he little realized that he was starting a new industry—the grape juice industry. At one time, there was a movement to have grape juice replace alcohol.

FOOD KITS

WELCH FOODS, INC.
Westfield, N.Y. 14787

WINE BY WELCH'S

Welch's is a name that we all grew up with—famous for its juice and jams and jellies. But did you know that Welch's also offers an extraordinary wine-making kit?

If you like grape juice, you'll love this wine. Naturally sweet, this kit provides you with everything you'll need to make ten bottles of wine, including your own labels! This is an easy wine to make, so if this is your first experiment in wine making at home, this might be just right for you. $9.95.

WELCH FOODS, INC.
Westfield, N.Y. 14787

MAKE YOUR OWN WELCH'S JELLY

It seems that kids like nothing better than a gooey peanut butter and jelly sandwich. Now you can provide them with that famous jelly, in homemade form. That's because Welch's puts out their own home jelly-making kit. It provides you with everything you need to make a dozen jars of jelly, including the jars and caps, the fruit juice base, the flavor enhancer, pectin, paraffin, and an information booklet. Not only can you choose from traditional Concord grape, but you can also ask for apple/crab apple and American cherry. The kit is only $8.99.

BACCHANALIA
273 Riverside Avenue
Westport, Conn. 06880

THE SAMPLER KIT

If you've ever had a yen to really fill up your wine cellar, why not try making your own. Everything you need is in the Bacchanalia Sampler Kit. This convenient wine-making system contains all the equipment necessary to produce your own gallon supply of high-quality red or white wine. You'll receive concentrate, yeast, Campden tablets, 1-gallon plastic jug, fermentation lock and cork, siphon hose, bottling corks, and decorated labels. And the Sampler Kit comes complete with a wine maker's book and complete instructions—to make wine making easy and fun. Cost is only $7.50—that's less than a bottle of good wine.

PAPRIKAS WEISS IMPORTER
1546 Second Avenue
New York, N.Y. 10028

BREAD SET KIT

Now you can have everything you need to guarantee successful bread baking with this eight-piece set of gadgets and aids. It includes a 12″ bread pan, a wooden spoon, a bread tester, a whisk, an 8-ounce measuring cup, a pastry brush, and a bread, biscuit, and roll recipe book. With all of these aids, your next bread-baking effort has to be a success! $19.98.

See International Groceries chapter for company description.

GREAT VALLEY MILLS
Route 309
Quakertown, Pa. 18951

MAKE YOUR OWN COTTAGE CHEESE!

Millions of Americans eat store-bought cottage cheese on a regular basis. If cottage cheese is your thing, you'll enjoy this new do-it-yourselfer: the Home Cottage Cheesery. This complete kit contains all the equipment you need to make many different kinds of cottage cheese. You'll receive a water-jacketed cheese vat, thermometer, instructions, and recipe booklets. Turn your kitchen into a fun food factory for the whole family. $11.95.

See Special Foods chapter for company description.

PUCKIN HUDDLE
Oliverea, N.Y. 12462

THE KIDS WILL THINK YOU'RE KOJAK

Kojak became famous with his Tootsie Roll lollipops—now you can become famous with your own homemade pops, too. Whoever heard of a lollipop kit? But there is one. You get six decorative molds—a heart, a soldier, a bunny, a four-leaf clover, an elephant, and a funny face, plus all six recipes and instructions for making your own homemade lollipops. What fun, what a delightful gift. $6.75.

PICKLE MASTERS
Box 65036
Los Angeles, Calif. 90065

MAKE YOUR OWN PICKLES

Surely, you must remember the vats of marvelous pickles you used to get from that favorite delicatessen in your home town. Nowadays, a good pickle is hard to come by. But you can make your own, right at home. All the ingredients you need to turn a cucumber into a pickle come in this kit. Not only can you pickle cucumbers, you can also pickle watermelon rind, squash, almost anything that moves you. The price is $10.00.

MANNA SUPPLY COMPANY
Box 3006
Santa Monica, Calif. 90403

GROW YOUR OWN SPROUTS

If you've ever been into health foods (or even Chinese foods, for that matter), you know all about the virtues of bean sprouts. Not only do they make a marvelous addition to salads, vegetables, and meat dishes, but they are filled with the B-complex vitamins, minerals, and protein. Even if you've had no success with growing anything before, it's guaranteed that you can grow bean sprouts. That's because they require no soil—only air and a little water. This kit provides you with everything you need to grow your own bean sprouts. And what's really fun is to bring a bag of fresh-grown bean sprouts as a hostess gift. The entire kit is $2.25 postpaid.

WALTER T. KELLEY CO.
Clarkson, Ky. 42726

MAKE YOUR OWN HONEY

This is a little more complicated than the heading makes it sound, since you must keep your own bees! Walter Kelley offers you a complete kit to start making your own honey. (There is a gourmet kit which also contains honey-making bees and a beautiful queen.) The basic kit provides you with a ready-to-assemble hive with complete instructions. There are also bee gloves (so you don't get stung), comb foundation, and a booklet on how to raise bees and collect honey. The price for the beginner's kit is $33.00. Naturally, with bees, it is more expensive—$55.00.

WAGNER PRODUCTS
Riverview Drive
Hustisford, Wis. 53034

HOMEMADE SAUSAGE

If you've ever been intrigued with the possibility of making your own sausage, here's your opportunity. Wagner's offers you a sausage-making kit. Not only is it easy, but it's also a real money-saver over gourmet sausages. It's available in two versions, the basic sausage kit and the more expensive gourmet sausage kit. The difference is that the gourmet sausage kit contains a deluxe sausage stuffer instead of an economy sausage stuffer. Since you save $4.95, you might want to start off with the basic kit and see how you do. Basic kit—$10.95. Gourmet kit—$14.95.

OLD-FASHIONED, HOMEMADE COTTAGE CHEESE

With so many people watching their weight, cottage cheese is a very popular food. Not only is it delicious, but it is also low in calories. If you'd like to try making your own cottage cheese, you might very well be intrigued with this cheese kit. Cottage cheese, fortunately, is easier to make at home than many of the other cheeses. So, if you are a beginner, you might want to start here. This kit contains a cottage cheese vat, thermometer, complete instructions, and recipes. $9.95.

MAKING OTHER CHEESE AT HOME

If after you've had a chance to make the cottage cheese, and you are interested in moving on from there, Wagner's offers you other kits which are easy to use and can make some interesting cheeses. Their basic kit contains everything you need to make homemade cheese, including the press, thermometer, rennet, and coloring tablets. They also give you recipes and detailed instructions. Although you will need a few kitchen ingredients and utensils, this basic kit provides you with everything you need to experiment with homemade cheese. $7.95.

VINO CORP.
80 Commerce Street
Rochester, N.Y. 14623

BRANDIED TOPPING

A meal is never complete without dessert, and so, consequently, cooks are constantly looking for new and exciting ideas. Well, here is one. Make your own brandied fruit topping for your favorite ice cream or fresh fruits in season. This kit provides you with a special brandied fruit yeast, rum flavoring (add a little extra rum of your own for the real thing), a ½-gallon rumtopf jar, and a traditional rumtopf ladle. You'll also get a special recipe book for more ideas. $7.99.

ACCESSORIES
437 Hyde Street
San Francisco, Calif. 94101

THE BREAD THAT MADE SAN FRANCISCO FAMOUS

You may remember San Francisco for Chinatown, Fisherman's Wharf, and the cable cars running up and down the many hills. However, San Francisco is famous for much more than that. Among its traditions is the San Francisco sourdough bread. This crusty loaf bread is served in all of the best restaurants. Over the years, it has become so famous that it is now available in fine restaurants across the country.

And you can make the real thing, right in your own home. This sourdough kit provides you with yeast, a stoneware crock, and instructions to make a marvelous sourdough bread. $7.95.

SEMPLEX OF U.S.A.
4805 Lyndale Avenue North
Minneapolis, Minn. 55412

FOR THE ADVANCED WINE MAKER

Here is a wine-making kit which will allow you to really begin to experiment in your homemade wines. The starter kit will provide you with everything you need, with the exception of sugar and water, to make your own wine. But what is really intriguing is the catalog, which offers dozens of other concentrates, yeasts, a vari-

FOOD KITS

ety of sugars, champagne corks, etc., to help you in the making of other wines. With all of this available to you, you'll be able to make some highly unusual wines. The starter kit sells for $11.95 postpaid.

PRODUCT SPECIALTIES
900 Jorie Boulevard
Oak Brook, Ill. 60521

STOCK YOUR OWN HOMEMADE WINES

If you have ever been interested in making your own wines, but thought that perhaps it was too complicated—well, guess again. Now you can make your own wine in a simple, two-step process. There is no siphoning, no straining, no filtering, no bottling, no sterilization, and no scrubbing. And you have a good selection of varieties—chablis, chianti, burgundy, rosé, and sherry.

This wonderful kit comes packaged in an unusual manner. The kit is made to look like a barrel. Not only do you "manufacture" your wine in this barrel, but it also becomes the serving method, too. The complete kit includes everything you need for simple wine making: yeast, yeast food, imported concentrate, wine skin and spigot, barrel, clarifier settler, air-lock assembly, and easy-to-read instructions. $11.99.

SWISS AMERICAN IMPORTING CO.
4354 Clayton Avenue
St. Louis, Mo. 63110

CHEESE BALL KIT

If you're a lover of fancy cheeses, here's a new way to make your own: the Dutch Garden Cheese Ball Kit. This complete, compact kit contains all the ingredients necessary to turn high-quality foods into delicious gourmet hors d'oeuvres. With its special formula cold-pack cheese food and distinctive blend of cashews, walnuts, and filberts, this easy-to-use kit makes every get-together a sensation. Use it the next time you throw a birthday party, give a baby shower, have friends over for cocktails, or prepare an elegant dinner. Whenever you use it, the Cheese Ball Kit is sure to add an exciting and original touch to your social occasions. $2.05.

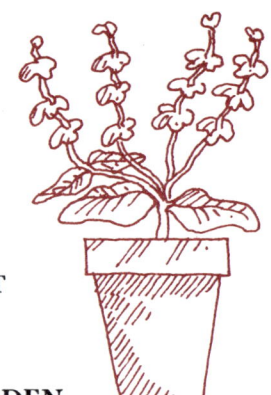

LE JARDIN DU GOURMET
Raymond Saufroy, Imports
West Danville, Vt. 05873

GOURMET HERB GARDEN

The Gourmet Herb Garden is much more than just a seed-starting kit. It's a whole hobby in a kit! It includes 24 Jiffy Pots, Jiffy Mix planting mix, plastic growing trays, seed germinating bags, eight varieties of herb seed (dill, oregano, parsley, basil, chives, marjoram, thyme, and sage), plus the "Gourmet Garden Guide," a 16-page booklet of easy-to-follow, kitchen-tested recipes and suggestions for using herbs in cooking. $4.95.

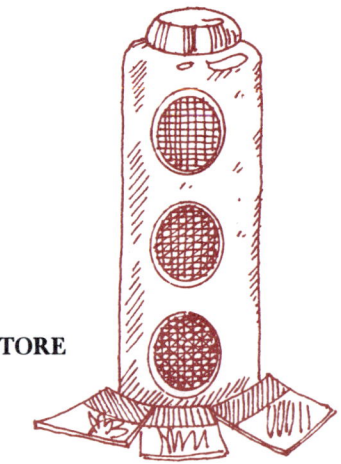

MOTHER'S GENERAL STORE
P.O. Box 506
Flat Rock, N.C. 28731

SPROUT-EASE

At last . . . the ideal way to produce sprouts! All you need is a standard, wide-mouth canning (or coffee or mayonnaise) jar and the Sprout-Ease Kit. The kit consists of three jar rings, each holding a different size wire-mesh screen and three sample seed packets. The fine screen is used for starting alfalfa, the medium screen for beginning mung beans or flushing away hulls from alfalfa sprouts, and the coarse screen for larger seeds or washing hulls from sprouted mung beans. The mesh rings are also available without the starter seed packets. Sprout-Ease Kit—$1.98. Sprout-Ease caps only—$1.39.

FOOD KITS

ABOUT THIS COMPANY

Mother's General Store offers you a variety of kits, which include yoghurt making, wine making, beer making, and a sourdough starter. Send $.25 for their 84-page catalog.

MOTHER'S GENERAL STORE
P.O. Box 506
Flat Rock, N.C. 28731

ROOT BEER EXTRACT

Here's the hard-to-find beverage base that was so popular before artificial carbonation put the bottling of soft drinks almost entirely in the hands of the large corporations. Once again, you can mix up a taste-tempting batch of root beer right at home with common yeast and sugar. One small bottle makes five full gallons and the extract keeps indefinitely so you can make split batches, too. Root beer extract—$.39.

See company description in this chapter.

OTHER SOURCES FOR FOOD KITS

MEADOWBROOK HERB GARDEN
Wyoming, R.I. 02898

JAMES G. GILL COMPANY, INC.
204–210 West 22nd Street
Norfolk, Va. 23517

BARREL WINERY
1201 University Avenue
Berkeley, Calif. 94702

PFAELZER BROTHERS
4501 West District Boulevard
Chicago, Ill. 60632

FARMER SEED AND NURSERY COMPANY
Faribault, Minn. 55021

NICHOLS GARDEN NURSERY
1190 North Pacific Highway
Albany, Oreg. 97321

J. T. MCCARTHY CO.
P.O. Box 7012
Milwaukee, Wis. 53213